Gender and the Science of Difference

Studies in Modern Science, Technology, and the Environment
Edited by Mark A. Largent

The increasing importance of science over the past 150 years—and with it the increasing social, political, and economic authority vested in scientists and engineers—established both scientific research and technological innovations as vital components of modern culture. Studies in Modern Science, Technology, and the Environment is a collection of books that focuses on humanistic and social science inquiries into the social and political implications of science and technology and their impacts on communities, environments, and cultural movements worldwide.

Mark R. Finlay, *Growing American Rubber: Strategic Plants and the Politics of National Security*
Jill A. Fisher, ed., *Gender and the Science of Difference: Cultural Politics of Contemporary Science and Medicine*
Gordon Patterson, *The Mosquito Crusades: A History of the American Anti-Mosquito Movement from the Reed Commission to the First Earth Day*
Jeremy Vetter, *Knowing Global Environments: New Historical Perspectives on the Field Sciences*

Gender and the Science of Difference

Cultural Politics of Contemporary Science and Medicine

EDITED BY
JILL A. FISHER

RUTGERS UNIVERSITY PRESS
NEW BRUNSWICK, NEW JERSEY, AND LONDON

LIBRARY OF CONGRESS CATALOGING-IN-PUBLICATION DATA

Gender and the science of difference : cultural politics of contemporary science and medicine / edited by Jill A. Fisher.

p. cm. — (Studies in modern science, technology, and the environment)

Includes bibliographical references and index.

ISBN 978-0-8135-5046-6 (hardcover : alk. paper) — ISBN 978-0-8135-5047-3 (pbk. : alk. paper)

1. Sex differences. 2. Sex differences—Political aspects. 3. Sex (Biology)—Social aspects. 4. Human biology—Philosophy. I. Fisher, Jill A., 1976–

QP81.5.G46 2011

612.6—dc22 2010040714

A British Cataloging-in-Publication record for this book is available from the British Library.

Visit our Web site: http://rutgerspress.rutgers.edu

Manufactured in the United States of America

CONTENTS

ILLUSTRATIONS

Figures

Tables

Gender and the Science of Difference

1

Gendering Science

Contextualizing Historical and Contemporary Pursuits of Difference

JILL A. FISHER

> Clearly, not all women and men are the same. But for many the science is undeniable: powerful hormones and the complex circuitry of the brain do shape our behavior and, therefore, our destiny.
>
> –ABC News, "The Truth behind Women's Brains"

> In analyzing male/female differences, these scientists peer through the prism of everyday culture, using the colors so separated to highlight their questions, design their experiments, and interpret their results. More often than not their hidden agendas, non-conscious and thus unarticulated, bear strong resemblances to broader social agendas.
>
> –Anne Fausto-Sterling, "Gender, Race, and Nation"

During the last ten years, there has been a resurgence of popular interest in the biological differences between women and men. Not surprisingly, given cultural obsessions with the brain, the primary site of these differences is often described as neurological. Books with titles such as *The Female Brain, The Essential Difference,* and *Why Men Don't Listen and Women Can't Read Maps,* among others, have been bestsellers as they promise readers knowledge about why men and women think and act differently. Computer metaphors are frequently mobilized so that differences between men's and women's brains are cast as "hardwired," communicating that they are both absolute and unalterable. Men and women are framed as inherently, biologically different, which is meant to explain and naturalize the social and behavioral differences between the sexes that the lay public so easily recounts. We are all familiar with these social and behavioral differences; they are inescapable in our culture: women are more emotional than are men, men are more sexually motivated than are women, women are listeners while men are problem-solvers, and so on and so forth. What is fascinating is that our society seeks to explain these differences as biological and that we invest vast scientific resources in "proving" that sex is the most powerful predictor of what type of individuals we will become.

Of course, from a scientific perspective, this is an overly simplistic rendering of the state of knowledge about the biological differences between the sexes. Nonetheless, science often plays it both ways. Professional scientists know that the biological *similarities* between men and women are greater than their differences even while they engage in research on those differences. They also know that a great deal of diversity exists within each sex. At the same time, however, scientific press releases by researchers' universities often exaggerate research findings in order to pitch to the media stories that will have broad popular appeal. In other words, while the media may be to blame for some of the representations of the biological differences between the sexes, many scientists are interested in and add to the popularization of such research findings. There are many reasons for this, but one among them is that scientists—those in the natural sciences as well as those in the behavioral and social sciences—want or need to illustrate the broader implications of their research in order to have successful careers.[1]

All of this is to say that it is important to think critically about scientific evidence—where it comes from, how it is generated, and why certain types of research are perceived as valuable. This means two things: first, science needs to be contextualized in a historical framework to understand how the pursuit of knowledge has changed and has remained the same over time and, second, contemporary research needs to be evaluated in the same way as historical examples of science to understand how current social and cultural norms continue to influence scientific inquiry. While the emphasis in this introductory chapter is primarily on the former, the case studies in the rest of the volume concentrate on the latter. Nonetheless, as I will discuss, there is a great deal of continuity between the past and the present.

Science as a Social and Cultural Enterprise

There are multiple ways to think about science and the production of knowledge. One element of science that is often overlooked is that in addition to being a way of understanding the natural world, science is also an institution comprised of individuals who are making decisions about what questions are important, what methods are valid, and how results will be analyzed. In popular culture, science is usually depicted as the process of revealing preexisting Truths about the natural world.[2] Yet facts are produced by individuals who use tools and methods that structure what claims can be made. In other words, as scientist Ruth Hubbard reminds us, "Facts aren't just out there. Every fact has a factor, a maker . . . [and] making facts is a social enterprise" (1988, 5). This is not to say that these scientific facts don't represent truths about nature; rather, it means that the facts that are created are partial and contingent interpretations of what humans observe. As will become more evident with several examples in the following sections, what scientists observe is often subject to what they already believe is true and is usually in sync with broader society's culture and values.[3]

Just as science is influenced by society, so too does science have an impact on how society is organized and functions. This reciprocal relationship illustrates that science is not a neutral enterprise. For example, Western society has long been characterized as patriarchal, meaning that it is hierarchically organized with men having the dominant position and women the subordinate one. On the ordinary level, this is witnessed through naming norms—children usually have their father's names and women typically take on their husbands' family names.[4] More than just customs, however, patriarchy is largely about power and control. It shapes what roles men and women have in society and our understanding of what each is capable of and what behavior is expected of each.[5] Patriarchal ideas have heavily shaped the pursuit of science, and science has supported the patriarchal system by naturalizing its norms and values. Many scientific pursuits can be labeled androcentric, meaning that knowledge has been constructed with a decidedly male focus or perspective. For instance, in studies of sexuality during the late nineteenth and most of the twentieth centuries, scientists focused their studies of sexual behavior on male animals, presuming that the male of the species (usually mice or rats) had the active role and that the females were passive, a view that mirrored contemporaneous presumptions about human sexuality (see Birke, this volume). It was only when women started entering the laboratory as scientists during the 1970s that female sexual behavior became a focus of investigation and the complexity of female behavior began to be observed (Van den Wijngaard 1995). In short, science tends to support the mainstream values of society, even when those values are sexist.[6]

So how then do we separate out *real,* bona fide differences from sexism or patriarchal values? Some would say that there are empirical differences between men and women that are simply biological and natural and, as a result, are completely apart from culture, politics, and so on. Many immediately argue that the main difference between women and men is that women can have babies while men cannot. Yet, practically speaking, the ability to develop a fetus and give birth is not what makes a person a woman; we wouldn't assert that someone is not a woman because she is infertile. Others point to external genitalia as the clear way to distinguish women from men. While breasts and labia and penises and testicles provide one means of categorizing people, there are, nonetheless, frequent examples of ambiguous genitalia as well as a great deal of variation in size and shape of both men's and women's parts. Adding to the slipperiness of these seemingly absolute categories of women and men, U.S. society is increasingly more sensitive to the experiences of transgender and transsexual individuals whose identities challenge strict binaries. Despite exceptions that are made, it is rare that we question how absolute our taken-for-granted assumptions about sex differences actually are and—more importantly—why ambiguity in these differences can make us (collectively speaking) nervous. Our cultural investment in these categories is one indication that interest in differences is never simply casual. Moreover, when we make distinctions between men and women, we are generally ascribing *meaning* to any differences identified between the groups.

As if physiological differences weren't tricky enough, the politics of the differences between the sexes comes most easily into view when science investigates behavioral patterns that diverge in women and men. These are the differences that are more accurately described as those of *gender* rather than sex, and the controversy lies in the degree to which gendered behaviors are "natural" or inherent to the sexes.[7] For example, women as a group are better nurturers than men are as a group. The question remains, however, whether nurturing behaviors are inherent to women or they are the product of socialization.

In order to make meaning out of documented or perceived gender differences, there are two predominant theoretical positions: *essentialism* and *constructivism*. Essentialism—also known as biological determinism—promotes the view that sex characteristics are natural and determine behavior. From this position, sex and gender are interchangeable concepts and terms. In contrast, constructivism denotes the view that sex has a biological basis while gender is socially constructed and not natural. This latter position arose from the women's movement of the 1960s, when feminists argued that gender differences have historically been mobilized to create and enforce inequalities between men and women in society, but gender norms can be redefined to equalize the sexes.[8] Or to put the two positions in more familiar terms, this is the old "nature versus nurture" debate using more precise terminology.

While many scientists borrow from both ends of the spectrum, there are radical examples of each of these positions. For example, scientists who work in the subdiscipline of sociobiology are known to make extreme claims leveraging evolutionary explanations about the biological basis of gendered behavior. Research in this field seeks genetic answers for why men supposedly prefer young, blond, large-breasted women (i.e., these characteristics are said to be perceived as indicators of women's fertility). On the other end of the spectrum are scientists who claim that even biological differences between the sexes are a function of social norms and customs. For example, studies show that exercise has more of an impact on physiology (e.g., bone density, muscle mass, metabolic rate) than does sex so that presumed sex differences are quite often not accurate in and of themselves but are instead proxies for differences in amount and types of physical activity (Birke 2001).

There might be evidence that different scientists can drum up to support both positions, but, generally speaking, mainstream Western culture tends to fall closer to the essentialism side of the spectrum. One bit of evidence for this is that as a society, we don't have many qualms about making claims about how men and women are inherently different. Even in a political climate in which people are concerned about being politically correct, it is rarely perceived as wrong to assert that women are inherently not as good at math and science as are men.[9] It should be no surprise then that much science ends up drawing conclusions that are reflective of society's broader beliefs; scientists are part of society, after all.

Does this mean that science is doomed to be biased? While there certainly is science that is biased, most science is conducted in ways that try to rid it of bias. The scientific method is one way to try to structure research so that it is done systematically and conscientiously. While most science is not biased, as such, all science is *interested*. This means that individual scientists, research communities, or nations are invested (and investing) in the research questions. Scientific projects are pursued because they are perceived as important, and this directs what types of knowledge about the world are generated. Good science—in the sense that researchers are following the standards of their field—is designed to prevent that interest from unduly influencing the results of the inquiry.[10] In spite of these efforts, it is impossible for individual researchers to cast off completely their personal and cultural values or even their economic or professional motives. Biologist Anne Fausto-Sterling refers to this as scientists' "blind spots." She explains that "a scientist may fail to see something that is right under his or her nose because currently accepted theory cannot account for the observation" (1992, 10). Research on sex or gender differences is especially complicated when most of the ideas people have about sex and gender tend to be seen as natural and are rarely interrogated. Science, therefore, may not be biased per se, but it is influenced by and contributes to patriarchy. One important goal is to examine how it has done so historically so that we can be more adept at detecting it in contemporary research.

Historical Trajectory of the Science of Difference

The pursuit of knowledge about biological differences between the sexes has a very long history. Although techniques and methods have changed considerably over time, questions about how and why women and men are different have occupied humans over the centuries from ancient philosophers and

Renaissance anatomists to modern-day geneticists. Interestingly, over time, the degree to which women and men have been perceived as different has increased even as more fine-grained details about the biological similarities between the sexes have accrued. The dominant ways of understanding similarities and differences have reflected both the capabilities of science to probe the human body and the historical preoccupations with the roles of men and women in their cultural contexts.

Throughout most of human history, the prevailing view of the sexes was premised on a "one-sex" model, in which females were thought of as imperfect or defective males (Laqueur 1987). Men's and women's reproductive anatomy was perceived as composed of identical structures with women's internal and men's external to the body. Anatomical renderings clearly depicted the organs as identical with the penis and testicles providing the model shape and structure for the vagina and uterus (fig. 1.1). The female body was viewed as inferior because it was thought to have failed in its development to the male form. Specifically, ancient philosophers and anatomists believed that proper human development was dependent on heat: the reproductive organs could only be pushed out of the body with sufficient heat in the developing fetus.[11] Females were understood to be the result of insufficient heat (a "mutilated male") and thus the imperfect sex (Tuana 1989).[12]

Aristotle had a profound influence on knowledge about human anatomy and reproduction, with many components of his model providing the foundation for beliefs about the sexes that went virtually unchallenged from the fourth century B.C.E. until the seventeenth century. In his view, variation in heat between the sexes could explain not only the physical differences in the reproductive organs but also that women are smaller and less muscular than are men. Aristotle further asserted that owing to less heat, women's brains are less developed, which makes women less intelligent and of poorer temperament. In her analysis of Aristotle's theories, philosopher Nancy Tuana framed and quoted his position: "And woman's inferior brain size in turn accounts for much of her defective nature. Woman is 'more jealous, more querulous, more apt to scold and to strike. She is, furthermore, more prone to despondency and less hopeful than the man, more void of shame, more false of speech, more deceptive, and of more retentive memory' " (1989, 148).

As evidence of men's greater heat, Aristotle claimed that male fetuses develop more quickly than female ones and that the greater number of male babies born with birth defects is caused by an excess in heat in those developing fetuses. At the same time, however, Aristotle set himself up for a problem. If heat speeds development, how then does his theory explain why girls reach puberty faster than do boys? He circularly explains that because females are colder they need less time to reach their own (inferior) perfection, sexual maturity. Tuana explains that "we can see from such inconsistencies in Aristotle's

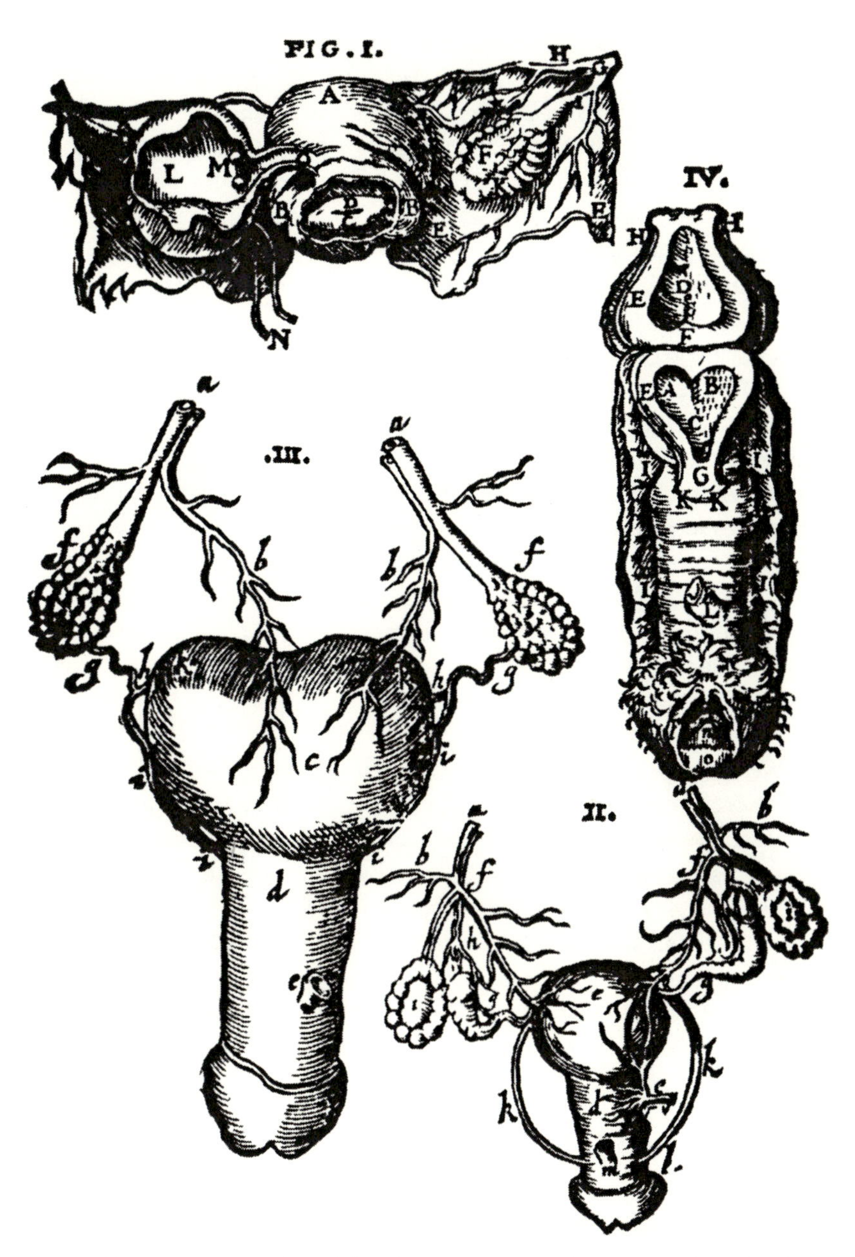

FIGURE 1.1 Representation of female anatomy as an internal version of male anatomy.

Source: N. Tuana, "The Weaker Seed: The Sexist Bias of Reproductive Theory," in *Feminism and Science*, edited by N. Tuana (Bloomington: Indiana University Press, 1989), 155.

theory that the doctrine that the female sex was inferior to the male was not a premise to be proved or justified, but was rather an implicit belief underlying Aristotle's development of his biological theory" (1989, 153).

Aristotle left it to his successors to explain why females have less heat than do males. Several hundred years later, in the second century, Galen "solved" the riddle. Through a creative representation of anatomy, Galen explained that it was the configuration of veins and arteries that control the amount of heat given to the developing fetus. Given cultural conceptions associated with the right (good) and left (sinister) during his time, Galen claimed that heat is associated with the flow of "pure" blood through the arteries on the right side of the body and that the lack of heat is a result of "impure" blood in veins on the left side of the body. The side of the father's body from which the "seed" came then determined the sex of the fetus. For instance, seed originating from the right testicle would by consequence lead to a male offspring, and seed originating from the left testicle would become female offspring. Galen's explanation is fascinating because it leverages a sophisticated understanding of the role of veins and arteries for delivering blood while being physiologically inaccurate: veins and arteries are not each located only on one side of the body. In spite of this amazing technical flaw, Galen's anatomical model continued to be the accepted explanation for sex differences for hundreds of years (until the seventeenth century). Tuana provides an explanation for the longevity of this false belief in spite of the evidence to the contrary:

> Anatomists persistently held to the view that the female seed was defective because of the impurity of the blood that fed it. Although careful attention to the actual structure of the veins and arteries of the testicles and ovaries would refute this view, anatomists continued to overlook this error. . . . It is perhaps not surprising that even an anatomist as careful as Vesalius would perpetuate such an error. The scientific theory he had inherited demanded this "fact." The belief that female seed arose from the "serous, salty, and acrid" blood of the left testes was the only viable explanation of the perceived differences between women and men. (1989, 161)

Science and technology studies (STS) scholar Nelly Oudshoorn (1994) has explained in her analysis of the gendered nature of science that these failures of observation were part of the dominance of the "one-sex" model. Anatomists were actively dissecting male and female bodies during the centuries, but they were not looking for differences between men and women or to challenge dominant theories of reproduction. Oudshoorn states, "The stress on similarities, representing the female body as just a gradation of one basic male type, was inextricably intertwined with patriarchal thinking, reflecting the values of an overwhelmingly male public world in which 'man is the measure of all things, and woman does not exist as an ontologically distinct category'" (1994, 6).

Supporting this point, Tuana (1989) notes that several anatomists documented inconsistencies between the standard physiological model of the time and their own observations, but they did not seem to understand the implications of their observations. In other words, it was not ignorance of anatomy that contributed to these scientific views but the cultural beliefs of the time about the relationship of men and women being premised on hierarchy, not difference.

In spite of the power of patriarchal thinking, Galen's model could not be sustained forever. With the spirit of experimentation as part of the scientific revolution, tools were developed to extend scientists' powers of observation. The invention of calculating machines, measuring devices, telescopes, microscopes, and so on changed the way that or degree to which nature could be understood. As improvements were made to microscopes, the visibility of cells came into focus, so that spermatozoa could be observed for the first time during the seventeenth century.[13] This "discovery" catalyzed a new science of embryology that could document the mechanics of reproduction at the cellular level and led to contemporary understanding of the roles of men and women in reproduction. It was through these changes in the methods, techniques, and subsequent findings of scientific inquiry that by the eighteenth century the one-sex model was replaced with an alternate patriarchal version of two distinct sexes.

With the emphasis on difference becoming the dominant way of perceiving the sexes in the eighteenth century, scientists began exploring and documenting differences in men's and women's bones, blood vessels, hair, sweat, and brains (Schiebinger 1989). The study of difference between the sexes became utterly pervasive. As part of the process of making new types of distinctions between the sexes, radical naturalizations of *femininity* began to occur within science as well as society at large. For instance, the belief that much of women's behavior (i.e., that which distinguishes their roles in society from men's roles) is natural began to be explained in terms of biological or physiological functions. This view continues to shape scientific inquiry, but the claims have not been static. Specifically, the site of the "essence" of femininity has shifted over time from the womb (uterus) in the eighteenth century to the ovaries in the nineteenth century to hormones in the twentieth century to the brain in the twenty-first century (Oudshoorn 1994). Women continued to be perceived as inferior to men, but no longer because they were seen as imperfect males; rather, the multiple biological differences that were identified cast them as weaker, more vulnerable, and so on.

Two lines of investigation about the differences between women and men have especially occupied scientists for the past two centuries: intelligence and sexuality. In examining the history of intelligence research, it is hardly an exaggeration to claim that scientists have sought measures of intelligence that yield the results they know *a priori* to be "true." In the nineteenth century, the newly established science of craniometry made inferences about intelligence by making measurements of the skull and brain. Women's brains were found to weigh on average five ounces less than men's brains.[14] The initial analyses performed were crude measures of absolute cranial capacity or brain weight. As philosopher Judith Genova explained, "Picturing the brain as a container and intelligence as some kind of vital fluid, [researchers] concluded that woman's smaller vessel could only hold less knowledge. Women were doomed to be less intelligent than men; their anatomy, wherever one looked, prevented it [from being otherwise]" (1988, 101).

This mode of measurement did not last for long when it became clear that other animals, such as elephants, have larger brains than do humans, and the idea that other animals could be more intelligent than humans was perceived as absurd or impossible. Scientists began experimenting with measurements: brain size in relation to body size, height, thigh-bone weight, and cranial height. Each of these measurements was in turn rejected because the results did not come out right: advantaging women in some cases or non-white peoples in other cases.

What guided the search for the "accurate" measurement was the certainty that the results would mirror the assumptions the researchers' held about the innate intelligence of the sexes or of racial and ethnic groups. These assumptions—or in some cases outright prejudices—were often right at the surface. Scientist Stephen Jay Gould conducted an examination of craniometry findings published in the nineteenth century and found a remarkable passage written by one of the leading scientists:

> "[Women's] inferiority is so obvious that no one can contest it for a moment; only its degree is worth discussion. All psychologists who have studied the intelligence of women, as well as poets and novelists, recognize today that they represent the most inferior forms of human evolution and that they are closer to children and savages than to an adult, civilized man. They excel in fickleness, inconstancy, absence of thought and logic, and incapacity to reason. Without doubt there exist some distinguished women, very superior to the average man, but they are as exceptional as the birth of any monstrosity, as, for example, of a gorilla with two heads; consequently, we may neglect them entirely." (Gustave Le Bon 1879, quoted in Gould 1992, 155)

Analysis of craniometry provides a perfect example of how closely linked social norms and cultural values are with scientific endeavors. Other types of research on intelligence have tended to exhibit far subtler manifestations of such bias.

Nonetheless, IQ and aptitude tests (like the SAT or GRE exams) have long been criticized for historically advantaging the affluent and white men while acting as a tool of discrimination against disenfranchised minority groups.

Twentieth-century research in psychology, neuroscience, and other related fields has likewise been used to "prove" that cognitive differences between men and women are inherent, naturalizing gender differences. Genova (1988) has explicitly associated the brain research of the 1970s and 1980s with craniometry because of its flagrant sexist agenda. The science being conducted during those decades focused in large part on hemispheric specialization and lateralization studies. Most of us know that research by the more colloquial terms of being "right brained" or "left brained." What is quite interesting about the history of this type of research is that the side of the brain that was assigned to each sex actually switched as expectations of gendered attributes and skills evolved during the same time period. Women were initially thought to be intuitive and holistic thinkers while men were said to be the logical and analytic ones. Today, however, women are said to be left-brain dominant, which means that they excel with verbal and analytic ability and process only one item at a time; men are said to be right-brain dominant, which means that they excel with spatial tasks, are more holistic, and can process multiple items simultaneously. In puzzling why the types of patterns of thinking that we attribute to the sexes would be reversed, it probably has to do with two trends that were occurring during the same time period: women began to enter the workforce in greater numbers as a result of the women's liberation movement and a new emphasis was being placed on creativity for professional and personal success. Genova writes,

> Once again men are viewed as natively equipped to do the truly inspired work. Women's cold, analytical, rational powers can only make them plodding amateurs in the creative game. Whatever the task, determinists will argue that men's holistic skills will assure them more efficient, more penetrating discoveries. The current attack on women, then, is confined to a kind of intelligence, it is specifically aimed at keeping them out of the world of science and trivializing their achievements in any field as routine and studied. . . . At last, it has been granted that women have the stuff, just not the right stuff. (1988, 103)

The perennial problem of research on difference is present here: the vast majority of the data supports the point that there is a great deal of overlap in how men's and women's brains function, but the differences command all the attention. Still, if one accepts that gender can be mapped onto different uses of the brain, the assumption that information processing is a proxy for intelligence is tenuous at best, especially when some types of highly valued activities require different types of information processing.[15] In spite of these obstacles to establishing the "truth" about the innate effects of sex on the brain, research on the gendered nature of cognition is still a major area of scholarship (see Wassmann, this volume).

Another area of science that has explicitly mobilized sex and gender differences has been research on sexuality. Just as the one-sex model presumed essential biological similarity between women and men, so too was there a belief that men's and women's sexuality was not so much divergent but different only by degree (men having more heat and all). At the same time that the one-sex model was replaced in the eighteenth century, opening up science to investigate the difference between men and women down to the cellular level, new norms and expectations began to govern women's sexual comportment. Historian Carol Groneman asserts,

> Well into the eighteenth century, both popular notions and medical understanding retained vestiges of the belief that women were as passionate, lewd, and lascivious as men were. . . . And yet by the nineteenth century, an ideology was firmly established: women by nature were less sexually desirous than men; the wifely and maternal role dominated their identity. . . . Profoundly suspicious of passion, Enlightenment and post-Revolutionary writers argued that women had less sexual desire than men and thus were uniquely suited to be a civilizing force; male passion would be controlled by the strength of woman's moral virtue. (1995, 225–226)[16]

Thus, women's sexuality and men's sexuality were placed on opposite ends of a spectrum.

Of course, individual women's experiences or expressions of sexuality did not necessarily reflect the cultural change. Medical science began to investigate women's deviations from the "norm" of feminine modesty, diagnosing the behavior as a new disease: nymphomania.[17] The women perceived to be most at risk were adolescents, especially blonds, and widows. As the type of women susceptible to nymphomania might suggest, the primary cure recommended to women was marriage. Should marriage not reduce women's "passions," physicians prescribed cold baths, vaginal borax douches, and abstention from red meats and alcohol. In cases of affluent women, especially wives caught masturbating or engaging in homosexual behavior, surgical options were presented, which included clitoridectomy (female circumcision) or oophorectomy (female castration through removal of the ovaries) (Barker-Benfield 1975). These more extreme interventions were justified by physicians because the medical belief was that nymphomania would lead to complete derangement of the mind and surgery on the genitals would help protect the brain from the disease.

The nineteenth-century pathologization and medicalization of sexuality was not confined to nymphomania. Scientific inquiry into the causes of and cures for homosexuality reflected the social and cultural angst about "contrary sexual instinct" when "homosexuals came to symbolize sterility, madness, and decadence in the late Victorian period" (Terry 1995, 132). During this time, homosexuality was seen as a constitutional predisposition, or, in other words, an *innate* characteristic that was simultaneously biological, psychological, and moral. There were competing explanations for why homosexuality occurred, including evolutionary causes, stress of industrial life, and gender confusion.[18]

As research on homosexuality continued into the first half of the twentieth century, explanations about its biological basis fell out of favor to be replaced by explanations about purely psychological causes. This shift in understanding about homosexuality made research into finding a treatment or cure for it more logical; it also led to fears that homosexuals were trying to "recruit" young men into their ranks, motivating some to want to identify, expose, and contain homosexuals. Cultural studies scholar Jennifer Terry describes the details of the Sex Variant study from the 1930s to 1950s led by a gynecologist to "devise a checklist of visible characteristics that could assist physicians in identifying homosexuals" (1995, 139). The study subjected forty male and forty female homosexual volunteers to a battery of psychological and physical examinations with the idea that homosexual behavior would leave clear signs of deviance on the body. The results of the study were that homosexuality is more psychological than physical. The researchers were dispirited because they found no distinct signs of homosexuality on the body that could provide evidence to physicians seeking to diagnose it.[19]

Scientific consensus in the middle of the twentieth century about the psychological basis for homosexuality was temporary; researchers continued to seek biological explanations. Specifically, the search for a "gay" gene has been a major area of genetics research. Other areas of investigation include research into brain structure, cognitive abilities, pheromones, fingerprints, length and girth of various body parts (e.g., limbs, feet, fingers, penis), and the auditory system (on the latter, see Spanier and Horowitz, this volume). From this list alone, there are unmistakable similarities between the search for a measure of homosexuality today with the search for a measure of intelligence in the nineteenth century. This type of scientific fishing expedition makes it clear that certain types of science are not just a phenomenon of the past, but that the same questions about difference preoccupy researchers in similar ways for similar purposes. The question then is not just how is science the same now as it is was in centuries past, but in what particular ways is it different today? And what do those differences suggest about our cultural moment?

Current Cultural Context of Science

Much rich scholarship has illustrated how past cultures of science were infused with patriarchal norms and values that influenced the kinds of research that was conducted and the interpretation of findings about differences between men and women. I have touched on only a few examples here. While the gendered values of past science may be easier for present-day observers to detect, it does not imply that current science is any less influenced by the broader social and cultural contexts in which contemporary research takes place.

Some argue that feminism has influenced culture and science to the point where women are no longer perceived as inferior (Bellafante 1998). Some even argue that, if anything, there is a tone of man bashing that might even imply that there is a social belief in *men's* inferiority. For instance, an ABC News *20/20* episode that aired in 2006 claiming to investigate, as titled, "The Truth Behind Women's Brains" seemed to send the message that women's brains just might be superior to men's brains:

> Girls . . . mature faster than boys, and girls' brains are as much as two years ahead during puberty. In fact, neuro-imaging shows that, early on, the typical teen girl has a stronger connection between the areas of the brain that control impulse—the amygdala—and judgment—the pre-frontal cortex. It may not be until late adolescence or their early twenties that boys' brains catch up to their girl peers. "To know that they're smarter than us by two years—it's a gap, it really is," said John Bessolo, one of the students in Dr. Brizendine's high school group. "They are the superior beings of the brain." (ABC News 2006)

Some commentators perceive this description of women's brain function as a feminist position and are encouraged by it.[20] They have argued that what is different about this neuroscience research from past scientific paradigms is that the research is not out to prove the inferiority of women and will likely be more reflective of "true" differences between the sexes. Even when women are cast as potentially *superior*, the hardwired "differences" are unmistakably gendered in highly problematic ways. As part of the *20/20* segment, effects of pregnancy and motherhood on the brain were highlighted:

> During pregnancy, these powerful hormones literally hijack a mother's brain circuits. She first becomes sleepy, hungry, and nauseous. Soon, the hormones oxytocin and prolactin intensely focus her maternal brain on the safety, and the needs of her child often to the detriment of everything else. . . . Triggered by hormones, a mother's brain becomes a virtual GPS systems [*sic*] for tracking and protecting her young. . . . A similarly dramatic hormonal effect is experienced when mothers breastfeed. As she nurses, oxytocin, the feel-good hormone, marinates a mother's brain. Many women say they are awash in feelings of warmth and pleasure. . . . For many women, their child-centric behavior not only compromises

their relationships, but also their jobs. Hormonally tethered to their child the interest of some mothers to return to work is often challenging. (ABC News 2006)

This hardly seems like an argument that will have positive implications for women because it normalizes traditional patriarchal assumptions that women's place is in the home, with children. It is also critical to note that research like this has been under fire. There have been numerous inquiries into the veracity of the findings being reported in news coverage like this as well as the research on which it is based (see Rogers, this volume).[21]

What does the popularity of this type of research about women's brains mean, then, for the relationship between science and cultural views of gender? Is this type of science merely an outlier that has received a lot of media attention? Or is it symptomatic of an underlying current in our culture? Both are probably true; it can be thought of best as representing the tension around women's issues in the United States as a result of the cultural context of postfeminism. *Postfeminism* is the term that refers to the view that feminism is obsolete because the battles it was fighting have been won. The postfeminist position is that in the past women did not have the same opportunities as did men, but because of the success of past women's movements, men and women are equal today. In the United States, there are very few employment opportunities from which women are explicitly denied, and women are excelling in college and entering graduate school in record numbers. Women today have many more choices than did generations before them.[22]

Postfeminism attempts to explain away any remnants of inequality between the sexes. For instance, the structural inequalities that persistently disadvantage women, such as the 20 percent less that women receive in pay than their male counterparts, are perceived as individual problems that require only individual solutions: if someone doesn't like her job, she should get another one; or women just don't choose to put in as many hours as do men. Postfeminism can also be thought of as an apolitical, have-it-all, or superwoman feminism: the job, the family, and the consumer products to stay young and attractive (Tasker and Negra 2007).[23] The absence of an analogue discourse of super*men* who also do it all, contributing at work and home, is either invisible or naturalized. This means that there is a paradox at the heart of postfeminism in which there is a simultaneous critique and reinforcement of sexism through a celebration of the choices women should and do have and an ignoring of the obligations from which men are excused.

If postfeminism is a major part of the cultural context of U.S. society today, what are its effects on science? This has yet to be documented in a systematic way, but the cultural politics of postfeminism could mean that there are fewer checks on scientific inquiry. For instance, it appears that it is no longer threatening to talk about the biological differences between the sexes because of the "equality" that has been achieved. Unlike in the 1960s and 1970s, there appears to be little controversial or offensive today about claiming that women are biologically fated to feminine gender roles. A belief that sexism has been eradicated likely means that we are not as attuned to perceiving sexism when and where it is mobilized. Moreover, it is unlikely that women entering the sciences today will have the same transformative effects on the scientific agenda as did the women who became scientists in the 1960s and 1970s.[24] Those early women scientists changed the questions that were being asked and the methods that were being used for collecting data. Women entering science in the 2000s and 2010s who subscribe to postfeminist views of society may be more likely to follow the male-dominated status quo in science than to offer new perspectives in their fields.[25] For instance, it may indeed be a symptom of postfeminism that no one stops to question the sexism inherent in research or the sexist implications of the findings. The purpose of this volume is to pause and consider how contemporary science mobilizes sex and gender in ways that reflect patriarchal—if not sexist—thinking about the roles of women and men in society.

Book Overview

This volume is organized into four parts that contain critical analyses of different aspects of contemporary science and medicine. Part 1, "Investigating Difference," includes three chapters that carefully examine claims that are being made by scientists today about the biological differences between men and women and between heterosexuals and homosexuals. The chapters tackle a range of scientific activity, representing research on genetics, neuroscience, and physiology. In her chapter "Sex Differences Are Not Hardwired," Lesley J. Rogers analyzes common types of explanations mobilized in the sciences for the causes of sex differences. She illustrates the flaws in some of these frameworks by refuting the scientific evidence or the conclusions drawn, and she argues for an alternative approach to understanding difference that embraces complexity in lieu of essentialism. For instance, she explores the literature documenting sex differences in infants' behavior to show that the socialization process begins at birth and cannot be dismissed as having no impact on sex differences. Bonnie B. Spanier and Jessica D. Horowitz question the conclusions that scientists make about the biological basis of homosexuality in "Looking for Difference? Methodology Is in the Eye of the Beholder." Drawing upon a case study of auditory research that purports to have found differences in ear emissions based on gender and sexual orientation, they offer a valuable mapping of the conceptual errors scientists are prone to make in their interpretations of data. Turning to brain imaging, Claudia Wassmann discusses research on sex-based differences in mathematical and cognitive abilities as well as emotional responses in "Evaluating Threat, Solving Mazes, and Having the Blues." She persuasively shows the ways in which brain-imaging studies reiterate and reinforce common gender stereotypes through the questions that researchers ask about differences in men's and women's brain functions.

Part 2, "Animal Obsessions," examines the ways in which human gender assumptions are inflected in research on nonhuman animals. The three chapters in this section include narratives of animal sexuality, but the analytic interest here is the attention that scientists and the media give to the behavior of nonhuman animals to represent taken-for-granted truths about human sexuality. In "Telling the Rat What to Do," Lynda Birke explains not only how researchers' assumptions about sex differences direct their hypotheses and observations, but also how laboratory animals' behavior and physiology are altered by their living conditions. She shows how purposeful breeding and laboratory contexts (such as being segregated by sex) tend to reinforce the assumptions that scientists have about animals' sex differences. Angela Willey and Sara Giordano's chapter, " 'Why Do Voles Fall in Love?' " returns our attention to research being conducting on the brain but with a twist. In their examination of monogamy gene research, they find "love" is a gendered story that gets written onto prairie vole behavior. In their fascinating account, they document how monogamy research positions only male behavior as being of interest because of assumptions that female monogamy is more "natural." In another case of concern with "aberrant" behavior, K. Smilla Ebeling and Bonnie B. Spanier explore popular interest in the question articulated in their chapter's title: "What Made Those Penguins Gay?" Zoos have long been sites of popular science and education through representations that they provide of the animal kingdom. Ebeling and Spanier trace the controversies and interventions that resulted from the politicization of media attention in the mid-2000s to "gay" penguins at German and U.S. zoos.

Part 3, "Categorizing Bodies," analyzes how scientific frameworks grapple with bodies that do not fit traditional sex and/or gender binaries. Both chapters in this section show how ambiguous bodies are defined and understood in ways that harm or disrespect the individuals who inhabit them. In "Intersex Treatment and the Promise of Trauma," Iain Morland describes the relationship between medical and psychological models of gender formation and the medical management of intersex. Exploring how knowledge about intersex is produced, he disturbingly illustrates the trauma caused to intersex individuals in order to minimize the parents' and physicians' cultural anxiety about children born with ambiguous genitalia. Sel J. Hwahng's chapter, "The Western 'Lesbian' Agenda and the Appropriation of Non-Western Transmasculine People," focuses on social science researchers' construction of knowledge about sex and gender through their treatment of anatomically female, transmasculine bodies. The chapter carefully documents the difficulty for researchers in thinking about and separating gender identities from biological bodies, especially when defining non-Western people through the lens of Western constructions of gender and sexuality.

Part 4, "Medical Interventions," examines the medical management of bodies that are constructed as "abnormal" as a result of gender expectations for those bodies. The three chapters in this section illustrate the medical imperative to intervene even when medical knowledge is absent, contradictory, or unfounded. In "Facial Feminization and the Theory of Facial Sex Difference," Heather Laine Talley investigates cosmetic surgery designed for male-to-female transsexuals. Analyzing the way the procedures are framed by surgeons as a way to correct "disfigured" faces, she reveals the process of "sexing" the face that occurs by defining what counts as "female" and "male" facial characteristics. Shirlene Badger's chapter, "The Proportions of Fat in Genetics of Obesity Research," concentrates on an entirely different context: a clinic conducting a study on the genetics of childhood obesity. Badger traces the scientific and personal narratives that are told to explain the causes of obesity, and she explores the gendered and familial meanings that parents and children in the study ascribe to fat. Finally, Emily Wentzell examines the medical management of erectile dysfunction (ED) in Mexico in "Making Male Sexuality." She describes how cultural and medical beliefs about gender and sexuality become fused in the treatment of ED in ways that differ unexpectedly from the medical model north of the border.

Together the essays in this volume depict contemporary science and medicine as gendered institutions that reflect society's cultural values. While there are doubtless many other areas of current scientific research that could be analyzed in this vein, this collection assembles examples that challenge the myth that contemporary science is value-free. More scholarship needs to be done to study explicitly the other types of cultural values at work in the modern research enterprise, such as how beliefs in ethnic/racial and class differences and hierarchies shape scientific practice. Nonetheless, scrutinizing scientific claims about the differences between women and men is an important place to begin because these assertions are especially prone to be mistaken for obvious truths.

NOTES

1. This is most obvious when it comes to securing funding for research. For example, the National Science Foundation, one of the U.S. federal government's primary sources of grant support for the sciences, requires scientists to include in their applications for research funds a statement detailing the "broader impacts" of the proposed research. The degree to which scientists can demonstrate the importance and relevance of their research will determine to some extent how successful they will be at receiving public funds.

2. Carolyn Merchant writes about the scientific revolution and the imagery connected with new methods of experimentation being used to uncover the secrets of nature: "In the seventeenth century . . . Francis Bacon (1561–1626) sets forth the need for prying into nature's nooks and crannies in searching out her secrets for human improvement" (1990, 33). These metaphors of scientific inquiry have contributed to the current sense that unknown facts are merely waiting for discovery.
3. There are, of course, examples in the history of science of dramatic departures from the status quo in observations, such as Copernicus's assertion in the sixteenth century that, contrary to the dominant belief at the time, it is the Earth that revolves around the Sun.
4. While it is becoming more common for women to retain their birth names when they marry, the vast majority of women still take their husband's last name. For example, one study estimated that more than 80 percent of U.S. women who are college graduates change their names when they marry (Goldin and Shim 2004), and the percentage of women who have not graduated from college who change their names is likely much larger.
5. What is important to understand about patriarchy is that it is a system, not just individuals. This means that it cannot be reduced to the people who participate in it. Because it is a system, it permeates our culture, is represented in our institutions, and structures our relationships. As a result, patriarchy is durable and tenacious even when individuals challenge the system (Johnson 2005).
6. While the focus of this volume is not explicitly on race or the construction of racial differences, science has also supported racist categorizations of different groups of people. Examples of science being used to "prove" racial inferiority can be found in American, European, and Japanese contexts and were often mobilized to justify slavery, colonization, or harsh treatment of ethnic minorities. See Fausto-Sterling 1995; Harding 1993; Terazawa 2005.
7. There are many ways to define gender. For the purposes of this discussion, gender can be thought of as the social practices that create and maintain distinctions between women and men. That these distinctions have been the fabric of inequality between men and women is not random or accidental, which is evinced by the way in which these distinctions become embedded in institutions and the rules of social interaction.
8. This idea gave rise to the "gender-neutral" childrearing movement, which focused on eliminating the gendered messages that children receive about differences between girls and boys and between women and men. The 1970s cult classic *Free to Be . . . You and Me* is an example of a television program and audio recording that challenged the traditional messages about gender that children receive.
9. There is one example where a claim of this sort quickly did become political. Former Harvard president (and economic advisor to U.S. president Barack Obama) Larry Summers made the remark in a speech that women do not have the same levels of scientific and mathematical aptitude as do men. The speech created quite the furor, leading to Summers's resignation from Harvard.
10. Defining "good science" can be a tricky endeavor because research can be technically correct but still problematic. For instance, clinical trials designed by the pharmaceutical industry to test their new products are good science in the sense that they follow the norms and standards of scientific practice, but studies are nonetheless designed so as to cast the drug in the best possible light (Fisher 2009). Sometimes it is not a

question of what science was done, but what science was undone, what knowledge was intentionally left unproduced (Hess 2007).

11. Up until the nineteenth century, science as we now would think of it was referred to as "natural philosophy." While early natural philosophers observed the physical world, it was not until the scientific revolution (beginning roughly in the sixteenth century) that experimentation with—or manipulation of—the physical world was incorporated into natural philosophy.
12. Nancy Tuana's 1989 piece on this topic ("The Weaker Seed: The Sexist Bias of Reproductive Theory") is a must-read for students interested in the history of science, and I draw extensively from its findings and analysis in this section.
13. Tuana (1989) relates a very interesting history of the perceptions of the roles of sperm and egg cells in contributing to reproduction. During the end of the seventeenth century and through the majority of the eighteenth century, scientists subscribed to the "preformation doctrine" of reproduction, which held that the organism was fully formed and had only to grow to become a person. Before sperm cells were visible under a microscope, scientists presumed that the egg must hold the preformed organism and that contact with male semen initiated its growth. This view was called "ovism." After sperm cells were made visible, some scientists then advocated that sperm held the preformed embryo, which could then grow upon being implanted in a woman's uterus. This view was called "animalculism." The latter view was more palatable to many at the time because ovism seemed to give too great a contribution to women in reproduction, but many fretted about the wastefulness of sperm containing preformed organisms.
14. Of course, the word "average" is critically important here. As with all averages, individual measurements deviate above and below the mean, and there is a fair amount of variability in the size of individual men's and women's brains. Thus, some women's brains weigh more than some men's brains. See Gould (1992) for a detailed analysis on why early measurements of men's and women's brains may have inadvertently introduced bias that accounted for the weight difference between the samples of men's and women's brains used.
15. Moreover, there is very little evidence to support that the sex differences found are innate differences. For instance, cross-cultural research often cannot replicate the findings of the Western world, and the brain is known to be a dynamic organ, not one that is fixed at birth. On this latter point, the brain has an amazing capacity to create new structures and patterns in response to injury or trauma.
16. Carol Groneman (1995) provides a fascinating discussion of how the rise of evangelical Christianity as well as the Industrial Revolution contributed to new mores around women's sexuality.
17. Sexual desire in men continued to be viewed as quite natural in the nineteenth and twentieth centuries. The term "satyriasis" was coined, however, to refer to the male analogue of nymphomania. The illness it represented was nonetheless believed to be much less common, less severe, and less damaging to men than nymphomania was to women (Groneman 1995).
18. Jennifer Terry (1995) describes three competing explanations for homosexuality that circulated in the nineteenth century. In the evolutionary explanation, researchers asserted that homosexuality was evidence of "overspecialization" of the species in which the highest levels of civilization were failing to procreate. This theory was based on the perceived overrepresentation of homosexuals in high society, especially those

who were intellectuals and artists. Other researchers claimed that modern industrial life put too many strains and constraints on individuals and that resulted for some people in the loss of adaptive ability, leading to homosexuality, here seen as a disorder of the brain and nervous system. A third line of research argued that homosexuality was evidence of a "third sex," one in which the female mind was in the male body and resulted in men behaving and appearing feminine. Likewise, lesbians were explained as a masculine mind present in a female body.

19. Alfred Kinsey's research on homosexuality during the same decades of the twentieth century even further raised alarm about the invisibility of homosexuals in society. Kinsey's controversial findings challenged the traditional binary between heterosexuality and homosexuality, suggesting that sexuality is instead a continuum and that people will engage in different types of sexual behaviors along that continuum over the course of their lifetime.
20. In fact, the *20/20* segment was highlighting the book *The Female Brain*, the author of which (Louann Brizendine) self-identifies as a feminist. There are, of course, many different kinds of feminism. Specifically, "difference" feminism—as opposed to liberal feminism—has as its basic tenet that women are naturally more nurturing than are men. It is a feminist position because people who subscribe to this view believe that women should not be disadvantaged or devalued by society for their difference. In the United States, where feminism is often thought of as a "bad" word, difference feminism does not have many subscribers, in part because people who might agree with this position would never identify as feminist. Perhaps Brizendine would categorize herself as this type of feminist.
21. Brizendine, author of *The Female Brain*, has been a particular target of attack as a result of her tendency to misrepresent or rely on seriously flawed science to make her claims.
22. Postfeminism is a cultural phenomenon; and, like much of our cultural beliefs, it is (problematically) based more on affluent and white women's experiences than those of poor women or women of color.
23. Empowerment rhetoric has been appropriated from the 1960s–1970s women's movements, so that empowerment is now seen as an individualized process, one that is best experienced through consumption (Fisher and Ronald 2008; McCaughey and French 2001).
24. In addition to the sexuality studies in which female rodent behavior became an object of scientific interest (Van den Wijngaard 1995), women entering science transformed primatology, biology, and medicine in the 1960s, 1970s, and 1980s (Haraway 1989; Schiebinger 2001).
25. I should clarify that not all women, and not all women entering science today, hold postfeminist views about society. Some of the science doctoral students that I have had the privilege of teaching have had incredibly strong views about the sexism and bias that can still be found in the sciences.

REFERENCES

ABC News. 2006. The truth behind women's brains. On *20/20*, September 28, 2006. Segment transcript available online: http://abcnews.go.com/2020/story?id=2504460 (accessed November 21, 2010).

Barker-Benfield, B. 1975. Sexual surgery in late-nineteenth-century America. *International Journal of Health Services* 5 (2): 279–298.

Bellafante, G. 1998. Feminism: It's all about me! *Time* 151 (25). Available online: http://www.time.com/time/magazine/article/0,9171,988616,00.html (accessed: January 14, 2010).

Birke, L. 2001. In pursuit of difference: Scientific studies of women and men. In *The Gender and Science Reader,* edited by M. Lederman and I. Bartsch, 309–322. New York: Routledge.

Brizendine, L. 2007. *The female brain.* London: Bantam Press.

Fausto-Sterling, A. 1992. *Myths of gender: Biological theories about women and men.* New York: Basic Books.

———. 1995. Gender, race, and nation: The comparative anatomy of "Hottentot" women in Europe, 1815–1817. In *Deviant bodies: Critical perspectives on difference in science and popular culture,* edited by J. Terry and J. Urla, 19–48. Bloomington: Indiana University Press.

Fisher, J. A. 2009. *Medical research for hire: The political economy of pharmaceutical clinical trials.* New Brunswick, NJ: Rutgers University Press.

Fisher, J. A., and L. Ronald. 2008. Direct-to-consumer responsibility: Medical neoliberalism in pharmaceutical advertising and drug development. In *Patients, consumers, and civil society,* vol. 10 of *Advances in Medical Sociology,* edited by S. M. Chambré and M. Goldner, 29–51. Brighton, UK: Emerald Publishing.

Genova, J. 1988. Women and the mismeasure of thought. *Hypatia* 3 (1): 101–117.

Goldin, C., and M. Shim. 2004. Making a name: Women's surnames at marriage and beyond. *Journal of Economic Perspectives* 18:143–160.

Gould, S. J. 1992. *The panda's thumb: More reflections in natural history.* New York: W. W. Norton.

Groneman, C. 1995. Nymphomania: The historical construction of female sexuality. In *Deviant bodies: Critical perspectives on difference in science and popular culture,* edited by J. Terry and J. Urla, 219–249. Bloomington: Indiana University Press.

Haraway, D. 1989. *Primate visions: Gender, race, and nature in the world of modern science.* New York: Routledge.

Harding, S., ed. 1993. *The "racial" economy of science: Toward a democratic future.* Bloomington: Indiana University Press.

Hess, D. J. 2007. *Alternative pathways in science and industry: Activism, innovation, and the environment in an era of globalization.* Cambridge, MA: MIT Press.

Hubbard, R. 1988. Science, facts, and feminism. *Hypatia* 3 (1): 5–17.

Johnson, A. G. 2005. *The gender knot: Unraveling our patriarchal legacy.* Philadelphia: Temple University Press.

Laqueur, T. 1987. Orgasm, generation, and the politics of reproductive biology. In *The making of the modern body: Sexuality and society in the nineteenth century,* edited by C. Gallagher and T. Laqueur, 1–41. Berkeley: University of California Press.

McCaughey, M., and C. French. 2001. Women's sex-toy parties: Technology, orgasm, and commodification. *Sexuality and Culture* 5 (3): 77–96.

Merchant, C. 1990. *The death of nature: Women, ecology, and the scientific revolution.* San Francisco: Harper & Row.

Oudshoorn, N. 1994. *Beyond the natural body: An archeology of sex hormones.* New York: Routledge.

Schiebinger, L. 1989. *The mind has no sex? Women in the origins of the scientific revolution.* Cambridge, MA: Harvard University Press.

———. 2001. *Has feminism changed science?* Cambridge, MA: Harvard University Press.

Tasker, Y., and D. Negra, eds. 2007. *Interrogating postfeminism: Gender and the politics of popular culture.* Durham, NC: Duke University Press.

Terazawa, Y. 2005. Racializing bodies through science in Meiji Japan: The rise of race-based research in gynecology. In *Building a modern Japan: Science, technology, and medicine in the Meiji era and beyond,* edited by M. Low, 83–102. New York: Palgrave.

Terry, J. 1995. Anxious slippages between "us" and "them": A brief history of the scientific search for homosexual bodies. In *Deviant bodies: Critical perspectives on difference in science and popular culture,* edited by J. Terry and J. Urla, 129–169. Bloomington: Indiana University Press.

Tuana, N. 1989. The weaker seed: The sexist bias of reproductive theory. In *Feminism and science,* edited by N. Tuana, 147–171. Bloomington: Indiana University Press.

Van den Wijngaard, M. 1995. The liberation of the female rodent. In *Reinventing biology: Respect for life and the creation of knowledge,* edited by L. Birke and R. Hubbard, 137–148. Bloomington: Indiana University Press.

PART ONE

Investigating Difference

2

Sex Differences Are Not Hardwired

LESLEY J. ROGERS

Although many examples of sex differences in behavior are exaggerations of minute and even trivial differences, there is no denying the existence of some sex differences in perception and cognition; and, therefore, it is important to ask what *causes* them. There are two radically different types of explanation of the cause: (1) unitary explanations and (2) interactive explanations. Unitary explanations claim that the differences are determined by the genes. In these explanations, higher-level accounts of sex differences in behavior and social position are reduced to accounts at the molecular level, and differences in behavior are said to be "hardwired," "a blueprint," "innate," or "essential." Genes, it is believed, have ultimate control either by acting directly to determine sex differences in brain structure and neural connections or by determining the levels of sex hormones, which, in turn, determine these differences.

By contrast, interactive explanations take experience (including learning and other cultural influences) into account and consider that during every stage of development contributions from experience, genes, and hormones interact in such complex ways that no one of these three sources of influence makes an overriding contribution in determining the sex differences in behavior.

Note that the debate about the cause of sex differences has moved away from the simple nature-versus-nurture dichotomy that used to be hotly debated some years ago. Interactive explanations do not deny contributing effects of the genes and hormones but take into account the fact that their expression is influenced by experience and that it is impossible to separate out any one of these influences from another (Bateson and Martin 1999, Rogers 2001; Rose 1997). Even those who claim that genes are the main cause of sex differences will often give lip service to a contribution of experience, although, ultimately, they see this as minor compared to the effect of the genes.

Genetic Determinist Explanations for Sex Differences

Genetic explanations for sex/gender differences in behavior are not at all new, but a recent upsurge of interest in them has taken place both within the realms of science and in the general population. In fact, there has been an apparently concerted effort to popularize genetic explanations in the media and in books for general readership, as evidenced by the popularity of Louann Brizendine's recent book (2007), *The Female Brain,* and by Simon Baron-Cohen's book (2003) *The Essential Difference: Men, Women, and the Extreme Brain.* These are just two of a large collection of books promoting unitary genetic and/or hormonal causation of the differences between women and men. For further examples see Deborah Blum (1997), who, although making reference to the effects of environment on gender differentiation and to the fact that gender differences are malleable, nevertheless leans heavily to the side of genes and hormones as determinants of the differences; and David Bainbridge, who believes that "men and women are *made* in fundamentally different ways" (2003, 33; emphasis mine, to point out that the assumption is one of essential differences that are hardwired). The simple message "genes cause behavior" may have nothing to do with the processes involved in differentiation of the behavior of women and men, which are much more complex, but it sells well in the media and it has sociopolitical power to maintain the status quo of gender roles, or even to return to earlier and more sharply defined role divisions.

Baron-Cohen (2003, 1) states from the outset that the female brain is "predominantly hard-wired for empathy" (seen as the ability to recognize the emotions expressed by others) and the male brain is "hard-wired for understanding and building systems." Brizendine follows this same line of thought but emphasizes the role of the sex hormones as mediators of the controlling genetic program. She sees these hormonal effects as "massive" (Brizendine 2007, 3) and believes that "the ability to connect deeply in friendship, a nearly psychic capacity to read faces and tone of voice for emotions and states of mind, and the ability to diffuse conflict . . . [are all] hard wired into the brains of women" (8). In fact, Brizendine even goes as far as saying that behavior, such as flirting, "is hardwired to testosterone" (2007, 89) the male sex hormone. It is not surprising, therefore, that she believes that sexual orientation is also "a matter of brain wiring" (86).

These genetic and hormonal explanations that either ignore the effects of experience or give it very little weight are part of a gathering wave of genetic explanations for human behavior in general. For example, Dean Hamer and Peter Copeland (1998) claim that genes determine intelligence, addiction to various drugs, core personality traits, homosexuality, even homelessness and, of course, sex differences in behavior. Hamer and Copeland are not alone in holding these views. Almost daily we hear of the discovery of a gene "for" this or that aspect of human behavior. Take, for example, the recent study claiming that a gene, sometimes referred to as "the warrior gene," on the X-chromosome is associated with gang membership and use of weapons (Beaver et al. 2010). As Steven Rose (1997) has pointed out, such reductionism is crudely ideological (see Rose 2009 and Stephen Ceci and Wendy Williams 2009 for recent debate on studying differences between racial groups in IQ).

Those who believe in genetic determinism see the human condition as unchanging and unchangeable. They give genetic reasons for the inequalities in society and thereby legitimize all manner of social oppression along racial, class, sexual, and other lines.

Claims that sex differences in behavior are hardwired deserve thorough critique because they have important consequences for women and men in society as well as in everyday aspects of each person's life. They are not confined to the popular literature but abound in the relatively recently emerging, and academically spurious, field of evolutionary psychology (for examples, see Buss 1995). Evolutionary psychologists believe that the behavior of modern humans still operates according to forces functioning in our evolutionary past. They allege that male-female differences were established by genetic selection during our hunter-gatherer past, when, according to them, men were the hunters and so needed to plan ahead, to form good mental maps of where they might find animals to hunt, to cooperate in hunting groups, and to develop good throwing ability (e.g., Buss 1995; Pinker 1994, 1997; Tooby and Cosmides 1990). These are all qualities that women then and now lack, according to the evolutionary psychologists. However, the assumption that the sex roles of hunter-gatherers were so strongly divided can be disputed; and the alleged labors of the women, which included finding vegetables, grain, and fruit, also required planning ahead and forming spatial and time-based, mental maps. In other words, firm evidence that different evolutionary processes of selection were operating in ancestral men and women is lacking.

There is even a claim that language appeared first in hominid males, due to a specific genetic mutation followed by a transposition of that mutated gene into the Y-chromosome (Crow 2002). Once in the Y-chromosome, it is said, this gene spread to all members of the human population (i.e., to both sexes) through genetic selection, said to have occurred because females found the males with the postulated "language" gene more attractive than those without it (Crow 2002). Not only is there no evidence of a single gene that might be responsible for the evolution of language (see Rogers 2004), but also the proposed male precedence in acquiring the postulated gene is pure fantasy with obvious sexism underlying the claim.

Added to this, there is the hypothesis that women are more monogamous and less competitive than men because they make a greater biological investment in the next generation (Trivers 1972; Wilson 1975). One step further, and according to Steven Pinker (1994), men banded together to fight tribal wars because victory led them to raping women, so maximizing their opportunities to pass on their genes to the next generation. This argument has been used, in present times, to justify raping of women by men.

Since neither brains nor behavior leave fossil records, all of these ideas of evolutionary psychologists are guess work. It is not difficult to refute them (see Kaplan and Rogers 2003) and show that they are a part of a conservative social force operating in support of patriarchal norms and values. They are used as political influence under the guise of being "science." We should be mindful of the fact that they have gained worldwide popularity and that many of their perpetrators hold positions of considerable status in leading academic institutions, predominantly in the United States.

Let us consider in more detail the main arguments put forward in support of the claim that genes determine male-female differences in behavior, and then the interactive roles of genes, hormones, and experience will be discussed, using some model biological systems to illustrate the influence of the sex chromosomes, sex hormones, and experience on the development of the brain and on the differentiation of behavior. The following describes four lines of evidence that are used frequently to substantiate claims that sex differences are determined by genes.

Early Age of First Appearance of Sex Differences in Behavior

If boys and girls show significant differences in behavior very early in life, it is often assumed that the differences must be determined by the expression of their XX and XY genomes, rather then being learned as a consequence of the different ways in which boys and girls are treated by their parents or other people who interact with them. The first fact to recognize here is that differential treatment of male and female babies begins from the day of birth (e.g., subtle differences in the way that the nursing staff and parents interact with the newborn child). Secondly, most studies of sex differences in early life test infants of around one to three years of age and still claim that any differences found must be caused by action of the genes alone, even though there has been ample time for cultural/experiential factors to have influenced the development of the children being tested.

Empirical evidence shows that adults actively reinforce sex stereotypes in infants in their first years of life. For example, an older but still important study by J. Ann Will, Patricia Self, and Nancy Datan (1976) tested the responses of women to children of about a year old when they were dressed at one time as a boy and another time as a girl. The women tested did not know the genetic sex of the children. When the child was dressed as a girl, they encouraged "her" to play with dolls and spoke to her gently. When he or she was dressed as a boy, they encouraged "him" to play with trucks and tools. Also, interactions with the child dressed as a boy were less verbal than they were with the child dressed as a girl. More recent research by C. Estelle Campenni (1999) has shown similar encouragement by mothers of sex-stereotyped play in boys and girls. Fathers also enforce sex stereotypes in their children. A cross-cultural study by Hugh Lytton and David Romney (1991) has revealed that parents are stricter with boys than girls and that fathers tend to differentiate between their sons and daughters more than do mothers. In fact, typically masculine behavior (as perceived according to conventional society's norms) in boys tends to be enforced by coercion (i.e., by punishment and force). Girls are encouraged, especially by their fathers, to adopt typical feminine behavior. It is not surprising, therefore, that boys show more agonistic behavior than girls at as early as two years old. Nor is it surprising that girls show a greater capacity than boys to empathize when they are only a year old and that girls have a "theory of mind" by three years old, whereas boys do not (reported by Baron-Cohen 2003). There has been sufficient time to learn these sex-typical characteristics under guidance by parents and other people with whom the child interacts.

Baron-Cohen (2003) suggests that parents (and other people) may treat their sons and daughters differently not because they (i.e., the parents) are behaving according to cultural norms, but because boys and girls *are,* in fact, different as a result of their genetic differences. In other words, according to Baron-Cohen and others, it is the genetic and hormonal differences between boys and girls that make adults treat boys and girls differently. However, this idea is not supported by the results obtained in the experiment mentioned above in which the same child was treated differently according to whether he or she was thought to be a boy or a girl and regardless of the child's actual genetic sex. It was the clothes that the child was wearing and nothing about its genes or hormones that altered the behavior of the adults. Hence, one can reason that, although the development of sex differences in behavior may, to some extent, involve interaction between genetically determined differences and differential treatment by adults, it is certainly not a process driven directly by the genes alone.

It is recognized that the closer to the time of birth that sex differences in behavior are found, the less likely that learning contributes to them. In this regard, it is especially relevant that one of Baron-Cohen's students has found sex differences in a sample of infants of just one to two days old (Connellan et al. 2001): the results showed that girls looked longer at a face than a mobile and the opposite was the case in boys. Although the difference in scores of looking time was not large, it would seem reasonable to assume that it is unlikely to be caused by different treatment of the girls versus boys on their first day of life. On the other hand, we cannot be entirely sure of this since seemingly small events, and events lasting for a very short time in early life, can have profound effects on the development of behavior. We know, for example, that young chicks form strong attachments to the hen or another attractive stimulus after a very short period of exposure to the hen, lasting for only minutes (Rogers 2010), and that maternal sheep imprint on the odor of their newly born lambs within minutes of giving birth (Kendrick et al. 1992). Hence, we cannot exclude the possibility that newborn humans are profoundly influenced by brief experiences in the first day of life.

We should also be aware of the fact that in modern Western societies the sex of the fetus is often known through ultrasound imaging. This opens the possibility for sex-differentiated experience prenatally via changes in behavior of the mother. Since there are examples of both physiology and behavior of young children being associated with stress experienced by the mother during pregnancy (e.g., de Bruijn et al. 2009), as shown also in animals (Weinstock 2008) and even carrying over two generations (Wehmer et al. 1970), it is possible that more subtle effects of maternal behavior during pregnancy might influence the development of sex differences in behavior. Therefore, even tests very close to the child's birth may not be free of sex-biased effects of experience.

Nevertheless, even if we accept Baron-Cohen's claim that the difference in attention to the face versus a mobile between one- or two-day-old girls and boys is due to genetic differences, we need to caution against extrapolating this finding to differences in behavior between adult women and men. Caution to the wind, Baron-Cohen claims that the difference is evidence that genes cause women to be "empathizers" and men to be "systemizers." This claim is based on the assumption that the face is a "social object" and the mobile a "physical-mechanical object" (Connellan et al. 2001).

What could be incorrect about this interpretation of the results? Firstly, if an infant looks longer at a face than at a mobile, that does not mean that the infant is *empathizing* with the face or that she perceives it as some kind of social stimulus: looking longer could simply mean the child has taken longer to habituate and that could be for precisely the opposite reason—because she has not been able to relate to it as a face or any other preconceived pattern. Secondly, since development is characterized by marked shifts in attention and behavior at different stages, sometimes changing from week to week, it is possible that the male infants might have found the face more interesting than the mobile if they had been tested when they were a little older, or that the opposite would occur in the female infants when they were a little older. Thirdly, it is highly unreliable to say that a fleeting measure of eye gazing preferences in newborn infants explains male-female divisions in employment patterns in adulthood, as Baron-Cohen (2003, 71) has suggested.

Our knowledge of the effects of experience on the behavior of very young human infants is extremely superficial, and, all too commonly, this gap in knowledge is filled by simplistic claims that sex differences are caused by the genes. As will be explained later, more detailed and controlled experimentation on the development of behavior in nonhuman species has revealed that the processes involved are far more complex than simple, reductionist causes in one direction from genes to behavior.

Measuring Brain Structure

In addition to sex differences in behavior, sex differences in various aspects of brain structure have been reported. For example, the ratio of gray to white matter is larger in women than men (Gur et al. 1999). Such differences in brain structure are often seen as proof that sex differences are hardwired. Quite incorrectly, we are prone to see differences in brain structure as being the cause of differences in behavior and pay no attention to the possibility that experience and behavior can themselves *cause* the structure of the brain to change.

Two lines of research operate in parallel these days. One is concerned with the plasticity of the brain, meaning that the brain can change its connections and even its gross structure in response to being exposed to stimulation or, indeed, to missing exposure to stimulation at particular stages of development (e.g., Rauschecker 1991). The other is concerned with describing differences in brain structure between different groups of people, sex differences included. Most of the research in the former area is conducted on animals because this allows use of techniques that cannot be applied to humans. Research taking the latter approach involves work on both animals (usually on rats) and humans. In fact, research on the human brain is growing rapidly as new techniques for imaging the human brain become available and are refined. So far, these two approaches to the study of the brain have usually been carried out by different laboratories, and different interpretations of the findings have been made with little integration of the ways of thinking.

A recent study by Jessica Wood, Dwayne Heitmiller, Nancy Andreasen, and Peg Nopoulos (2008) is worth discussing in order to illustrate a particular line of thinking by neuroanatomists studying sex differences. Comparing a sample of thirty women with thirty men, these researchers found that the size of the ventral prefrontal cortex region of the brain relative to the size of the whole brain was significantly larger in women than men. They chose to look at this region of the brain because it is said to be important in interpreting nonverbal, social cues and because there is some evidence that women and men differ in their attention to these cues. The size of this part of the brain was calculated from measurements of the length of the straight gyrus, a subregion of the ventral prefrontal cortex, and they found it to be larger in their female sample than in their male sample. They also found that the size of this region correlates positively with individual performance on a test of interpersonal awareness. The reductionist way of interpreting this result is to say that it provides evidence that women are hardwired to have a greater capacity for social awareness and to see this as coming about as a result of evolutionary selection dependent on the role of women in raising children. In fact, Wood, Heitmiller, and colleagues (2008) were rather less inclined to adopt this "hard-line" genetic interpretation of their results. Nevertheless, although they discussed potential interaction between experience and genetic influences, they concluded their paper by saying that there is a "strong genetic component" to the development of social cognition and, by implication, to the development of sex differences in the structure of the brain region that they studied.

A further study by the same researchers suggested that any unitary explanation for the cause of their results would be incorrect or, at least, vastly oversimplified because, when they examined the size of the same region of the brain in children, they found a result opposite to that found previously in the adults (Wood, Murko, et al. 2008): the region was larger in boys than in girls. Clearly, the structure of this region of the brain changes with maturation or experience, and likely with contributions from both of these factors. Interpretation of the functional meaning of a difference in brain size is much more complicated than the results that one stand-alone study might have suggested.

Furthermore, Wood, Heitmiller, and colleagues (2008) alerted us to another very important aspect of such research by giving their subjects a psychological test of "gender," which assessed the degree of masculinity versus femininity based on the subjects' interests, abilities, and personality. (Note that the gender score frequently does not match the biological sex.) The measure of gender correlated with the size of the ventral prefrontal cortex irrespective of the subjects' biological sex: higher femininity was matched with a larger size of this region. Therefore, the size of this region of the brain reflects gender rather than genetic/biological sex per se. Since gender is shaped by experience, it follows that the differences found in the structure of the brain are not simply hardwired.

Such research illustrates the need to move on from just looking for structural differences in the brains of women versus men, or girls versus boys, and to include assessments of gender. In addition, it should always be remembered that not only does brain structure and circuitry affect behavior, but behavior and experience also affect brain structure.

Absence of Variation across Cultures

If the same differences between women and men occur in different cultures, it is commonly assumed that they have been determined by evolutionary selection and are, therefore, encoded in the genes (e.g., Baron-Cohen 2003, 94; Brizendine 2007, 2). This is weak evidence, if it is any evidence at all, since it is equally possible that the differences in behavior are established and maintained by learning. Learned patterns of behavior can be as resistant to change as any behavior that might be encoded in the genes, and they can also be passed on from one generation to the next.

Also, invariance of male-female differences across cultures is often claimed on the basis of testing only a few groups from cultures not differing greatly from each other. In fact, only one exception to the pattern, in one culture, is sufficient to break the rule and put the argument for genetic causation into doubt. For example, the claimed cultural invariance of male superiority in spatial ability was undermined by a study by John Berry (1966) showing that Eskimo women are superior to men in spatial ability (see also Berry 1971). This single exception of the "rule" of male superiority in spatial perception undermines the evolutionary psychologists' claim that sexual selection taking place in ancestral humans explains male-female differences in spatial ability.

Existence of Sex Differences in Animals

The existence of sex differences in the behavior of animals is often taken as evidence that these differences are genetically determined. This belief relies on an incorrect assumption that nonhuman animals acquire their sex-typical behavior either entirely without learning or, if learning is involved, its role is minimal compared to the control by genes. Contrary to this view, elegant studies by Celia Moore in the 1980s (1984, 1985) showed just how important early experience is for the development of sex differences in the behavior of adult rats.

Moore and colleagues showed that mother rats lick the anogenital region of their pups frequently, and male pups are licked more than female pups (Moore 1984; Moore and Morelli 1979). This difference in stimulation leads to development of sex-typical differences in behavior, as shown by the fact that stimulating the anogenital region of female pups with a paintbrush leads them to show typical male behavior when they are adults. Also, no sex differences develop in litters that are raised by mothers that have had small tubes inserted into their nostrils to ensure that they cannot tell the difference between their male and female pups and so lick them equally. The tubes block the mothers' ability to smell, and this is the sense that she uses to tell the male from the female pups: male pups have higher levels of the hormone testosterone than do female pups, and this causes them to produce urine that attracts their mother to lick their anogenital region (Moore 1985).

The above experimental evidence is further supported by the fact that female pups injected with testosterone are licked as much as male pups, and they too develop male-typical behavior. Hence, behaving as an adult male or female does not depend on a direct action of testosterone on the pup's brain, as many have assumed, but on the interactive influences of genes, secretion of the hormone testosterone, and maternal treatment during early life. Early experience interacts with the effect of the genes (Moore et al. 1997), and does so in an entirely integrated way.

These detailed, controlled, and innovative experiments on rats have helped us to understand the complex interactive processes of development that lead to sex differences in behavior. They also caution us not to assume that hormone levels (e.g., testosterone) either before or after birth change behavior by acting directly on the brain. Although it is often assumed that in humans, as well as animals, testosterone acts directly on the brain to "masculinize" it (e.g., Brizendine 2007), the hormone's influence can be indirect, acting by changing the way parents treat their male and female offspring in some way. These ways of changing behavior are equally likely to be true for primates (Kaplan and Rogers 1999), including humans, as they are for rats, although the type of behavioral interaction that affects the development of sex differences would almost certainly vary from species to species.

It is often assumed that sex differences seen in monkeys and apes are similar to those seen in humans (e.g., Alexander et al. 2009). For example, Baron-Cohen believes that female monkeys, apes, and humans all show more interest in babies than do males and that this may be because females have heightened emotional sensitivity (2003, 96–97). Likewise, Brizendine draws attention to studies on macaque monkeys that show higher levels of rough-and-tumble play in juvenile males than females and uses this as evidence that sex differences in play by humans are innate (2007, 25). Such general statements about primates have little validity since attention to offspring varies greatly from one species of primate to another and even between individuals. In marmoset monkeys, for example, the male carries out much of the caring for offspring. In apes, as we know all too well from experience with animals in zoos, the female has to learn to be a parent. Moreover, citing the evidence for a sex difference in the play behavior of macaques is selective because, although infant male macaques show more rough-and-tumble play than females, this difference disappears as they get a little older.

Brizendine also highlights a postulated "monogamy gene" allegedly present in prairie voles that form monogamous pairs and not in montane voles that have multiple partners (2007, 73). She extrapolates this to humans to say that only some men have this "monogamy" gene and that they are the more desirable partners for women. Suffice it to say that this extrapolation of research on voles to humans bridges so many species that it is most unlikely to have any validity (see Willey and Giordano, this volume, for a more detailed discussion of voles).

Interactive Explanations for Sex Differences

Over recent years the focus on molecular genetics and the accompanying rise of sociobiology, followed by evolutionary psychology (Bateson 2005; Kaplan and Rogers 2003), has marginalized studies examining the interaction between genetic and epigenetic (experience) influences on the development of behavior. The central dogma of molecular biology is a unidirectional pathway of causation from genes through proteins to structure and behavior (discussed by Lewontin 1991). The supremacy of this view of biological causation has taken sway despite the importance of understanding the multiple and reciprocal influences between levels of organization (genetic, neural, environmental, meaning both behavioral and social, etc.) on an organism's development (Gottlieb and Lickliter 2007). Development of both physical form and behavior involves continual dynamic interactions between genes and environment (Gottlieb 1998, 2000).

Living systems can be described on many levels (Rose 1997). According to the traditional hierarchy of science from the most complex to the simplest, the range of levels is sociological, psychological, physiological, cellular, subcellular (genes), and biophysical. Bidirectional interactions take place both within and between levels. Some scientists studying the development of living organisms (e.g., Oyama 1985; Rogers 2001; Rose 1997) have drawn attention to the complex interactions between genetic and environmental influences and stressed the importance of moving away from reductionist thinking, meaning that the causation of complex behavior (sociological or psychological) should not be reduced to causal explanations at another, lower level (genes). Explanations at one level should not contradict explanations at another level, but those at a lower level cannot subsume those at a higher level. In other words, sex differences in genes do not provide an explanation for sex differences in behavior.

Studies of development in precocial animals have been crucial in illustrating that experience/learning in early life can have over-riding effects on behavior in later life; and, speaking generally, the findings apply to all vertebrate species, including humans (Bateson and Martin 1999). At particular stages of development, certain types of learning occur more readily and have longer-lasting effects than they do at any other stage. These stages are known as sensitive periods. There are many different kinds of sensitive periods, of different durations and at different times of life, when different developmental or learning effects take place, and one sensitive period may not be independent of the others (Rogers 2010). As Bateson and Martin (1999) have stated, each sensitive period represents a dynamic interplay between the individual organism's internal organization and the external conditions.

During sensitive periods the organism is open to specific experiences that change its neurochemistry and can change the structure of the brain in ways that have long-lasting and profound consequences on behavior (Bateson 1979). Each sensitive period remains open until the appropriate experience has occurred and then it closes (or passes). Should the appropriate experience not occur during the sensitive period, the sensitive period may remain open for longer than usual, but eventually it will close without the crucial developmental step being taken. This, too, has long-lasting consequences on both the brain and behavior.

An example of a sensitive period much studied is that of imprinting by precocial birds, such as ducklings and domestic chicks. Soon after it has hatched the young bird is attracted to any conspicuous visual stimulus, on which it will imprint. That stimulus is usually the mother duck or hen, but it can also be a red ball, a flashing light, or some other attractive visual stimulus. Once the young bird has imprinted, it will stay near that stimulus and follow it when it moves. This is an example of powerful learning that is not at all easy to reverse. Imprinting is associated with changes in the neural circuits in a particular region of the brain and changes in the levels of neurotransmitters. Such changes demonstrate how greatly the brain can change as a result of experience. If the young bird is kept in conditions in which there are no conspicuous visual stimuli (e.g., in darkness), the imprinting period is extended beyond the usual time; but by about the end of the first week of life it closes without the chick imprinting, which also has profound effects on later behavior.

From research on these species we know that sensitive period of imprinting is but one of a number of sensitive periods when different types of experience and learning have major effects on behavior (Rogers 2010). Some of these periods occur even before hatching: light exposure of the eggs just before hatching, for example, affects the development of the visual pathways and behavior in ways that are long-lasting and essential for survival (Rogers 1990).

Sensitive periods also occur in mammals, including humans, although we have far less precise knowledge of these, particularly in terms of the accompanying changes in the brain. We do know, however, that experience during one sensitive period affects experience in the next and so on. The processes of development are thus not linear from one event to the next but interactive, complex, and interwoven. Although the expression of genes may trigger the beginning of the process (possibly opening the first sensitive period or, in the case of chicks, determining what is an attractive visual stimulus), the genes have little or no ultimate role in guiding the process.

In other words, the X- and Y-chromosomes with their respective genes should not be seen as canalizing the development of males and females in tightly constrained and diverging paths. Wendy Wood and Alice Eagly (2002) see sex differences in behavior as arising from interactions between the social and physical environment and constraints imposed by genetically determined sex differences. The genetically determined differences to which they refer are reproduction and physical size and strength. Admittedly, their view is an advance on the unitary, reductionist explanations discussed earlier (see also Eagly and Wood 1999), but their concept of development is more simplistic than the one I am suggesting. In my opinion, the processes of brain and behavioral differentiation between the sexes are even more flexible and open ended.

While it may certainly be true that expression of the X- and Y-chromosomes has an effect on the brain before birth, such effects are far from proven to make any functional contribution to sex differences in behavior and far from known to impose any guiding influence on the development of sex differences in behavior. Some scientists hold out great hope for finding sex differences in the expression of genes in the human fetus, as seen by the recent publication of a paper by Björn Reinius and Elena Jazin (2009). These researchers looked at the expression of genes encoded in the Y-chromosome in various regions of the human brain before birth and claimed to have found a sex difference based on a comparison of three female fetal brains with just one male fetal brain. Clearly, with that sample size, their claimed finding is premature; but it was published, and they said that these genes are "likely having functional consequences for sex bias during human brain development" (Reinius and Jazin 2009, 988). Moreover, the media took it up as evidence of a sex difference without any mention of the number of subjects used in the study. Even if it is shown conclusively that cells with XX chromosomes express some different genes than cells with XY chromosomes, it will not tell us what contribution, if any, this makes to sex differences in behavior and cognition in children or adults, as the researchers also admit.

Elinor Ostrom, the woman awarded the 2009 Nobel Prize in economics, has pointed out that we must embrace complexity because simple models are worthless if they do not reflect reality (cited by Irvine 2009). Admittedly, she was referring to ways of achieving a socio-ecological sustainability and not to explanations for sex differences in behavior, but her sentiment applies to the latter also. It is time we moved away from static and simplistic views of brains and behavior.

In concluding, it is important to note that, as shown empirically by Janet Hyde (2005), the differences between the sexes are not as large as commonly implied by scientists and popular writers (e.g., Gray 1992, 2008, who vastly exaggerated the psychological differences between the sexes). The claimed differences do not often hold up from one study to the next; and, even when significant differences are found, they are often quite marginal (Hyde 2005; Halpern et al. 2005). The large overlap between the sexes is far too often ignored.

REFERENCES

Alexander, G. M., T. Wilcox, and R. Woods. 2009. Sex differences in infants' visual interest in toys. *Archives in Sexual Behavior* 38:427–433.

Bainbridge, D. 2003. *How the X chromosome controls our lives.* Cambridge, MA: Harvard University Press.

Baron-Cohen, S. 2003. *The essential difference: Men, women, and the extreme male brain.* London: Allen Lane.

Bateson, P.P.G. 1979. How do sensitive periods arise and what are they for? *Animal Behaviour* 27:470–486.

——. 2005. The return of the whole organism. *Journal of Bioscience* 30:31–39.

Bateson, P., and P. Martin. 1999. *Design for a life: How behaviour develops*. London: Jonathan Cape.

Beaver, K. M., M. DeLisi, M. G. Vaughn, and J. C. Barnes. 2010. Monoamine oxidase: A genotype is associated with gang membership and weapon use. *Comprehensive Psychiatry* 51:130–134.

Berry, J. W. 1966. Temne and Eskimo perceptual skills. *International Journal of Psychology* 1:207–229.

——. 1971. Ecological and social factors in spatial perceptual development. *Canadian Journal of Behavioural Science* 3:324–336.

Blum, D. 1997. *Sex on the brain: The biological differences between men and women*. New York: Viking.

Brizendine, L. 2007. *The female brain*. London: Bantam Press.

Buss, D. M. 1995. Evolutionary psychology: A new paradigm for psychological science. *Psychological Enquiry* 6:1–30.

Campenni, C. E. 1999. Gender stereotyping of children's toys: A comparison of parents and nonparents. *Sex Roles* 40:121–138.

Ceci, S., and W. M. Williams. 2009. Should scientists study race and IQ? Yes: The scientific truth must be pursued. *Nature* 457:788–789.

Connellan, J., S. Baron-Cohen, S. Wheelwright, A. Batki, and J. Ahluwalia. 2001. Sex differences in human neonatal social perception. *Infant Behavior and Development* 23:113–118.

Crow, T. J., ed. 2002. The speciation of modern *Homo sapiens*. *Proceedings of the British Academy*. Vol. 106. Oxford: Oxford University Press.

de Bruijn, A.T.C.E., H.J.A. van Bakel, H. Wijnen, V.J.M. Pop, and A. L. van Baar. 2009. Prenatal maternal emotional complaints are associated with cortisol responses in toddler and preschool age girls. *Developmental Psychobiology* 51:553–563.

Eagly, A. H., and W. Wood. 1999. The origins of sex differences in human behavior. *American Psychologist* 54:408–423.

Gottlieb, G. 1998. Normally occurring environmental and behavioural influences on gene activity: From central dogma to probabilistic epigenesis. *Psychological Review* 105:792–802.

——. 2000. Environmental and behavioral influences on gene activity. *Current Directions in Psychological Science* 9:93–97.

Gottlieb, G., and R. Lickliter. 2007. Probabilistic epigenesis. *Developmental Science* 10:1–11.

Gray, J. 1992. *Men are from Mars, Women are from Venus*. New York: HarperCollins.

——. 2008. *When Mars and Venus collide*. London: HarperCollins.

Gur, R. C., B. I. Turetsky, M. Matsui, M. Yan, W. Bilker, P. Hughett, and R. E. Gur. 1999. Sex differences in brain gray and white matter in healthy young adults: Correlations with cognitive performance. *Journal of Neuroscience* 19:4065–4072.

Halpern, D. F., C. P. Benbow, D. C. Geary, R. C. Gur, J. S. Hyde, and M. A. Gernsbacher. 2005. The science of sex differences in science and mathematics. *Psychological Science in the Public Interest* 8:1–51.

Hamer, D., and P. Copeland. 1998. *Living with our genes*. New York: Anchor Books.

Hyde, J. S. 2005. The gender similarities hypothesis. *American Psychologist* 60:581–592.

Irvine, J. 2009. Girl power takes on selfishness. *Sydney Morning Herald*, October 16:17.

Kaplan, G., and L. J. Rogers. 1999. Parental care in the common marmoset (*Callithrix jacchus jacchus*): Development and effect of anogenital licking on exploration. *Journal of Comparative Psychology* 113:269–276.

———. 2003. *Gene worship: Moving beyond the nature/nurture debate over genes, brain, and gender*. New York: Other Press.

Kendrick, K.M., F. Lévy, and E. B. Keverne. 1992. Changes in the sensory processing of olfactory signals induced by birth in sheep. *Science* 296:833–836.

Lewontin, R. C. 1991. *The doctrine of DNA*. London: Penguin Books.

Lytton, H., and D. M. Romney. 1991. Parents' differential socialization of boys and girls: A meta-analysis. *Psychological Bulletin* 109:267–296.

Moore, C. L. 1984. Maternal contributions to the development of masculine sexual behavior in laboratory rats. *Developmental Psychobiology* 17:347–356.

———. 1985. Sex differences in urinary odors produced by young laboratory rats. *Journal of Comparative Psychology* 99:336–341.

Moore, C. L., and G. A. Morelli. 1979. Mother rats interact differently with male and female offspring. *Journal of Comparative and Physiological Psychology* 93:677–684.

Moore, C. L., L. Wong, M. C. Daum, and O. U. Leclair. 1997. Mother-infant interactions in two strains of rats: Implications for dissociating mechanism and function of a maternal pattern. *Developmental Psychobiology* 30:301–312.

Oyama, S. 1985. *The ontogeny of information: Developmental systems and evolution*. Cambridge: Cambridge University Press.

Pinker, S. 1994. *The language instinct*. New York: William Morrow.

———. 1997. *How the mind works*. New York: W. W. Norton.

Rauschecker, J. P. 1991. Mechanisms of visual plasticity: Hebb synapses, NMDA receptors, and beyond. *Physiological Reviews* 71:587–615.

Reinius, B., and E. Jazin. 2009. Prenatal sex differences in the human brain. *Molecular Psychiatry* 14:988–989.

Rogers, L. J. 1990. Light input and the reversal of functional lateralization in the chicken brain. *Behavioural Brain Research* 38:211–221.

———. 2001. *Sexing the brain*. New York: Columbia University Press.

———. 2004. Leaping to laterality and language: The case against. *Laterality* 9:225–232.

———. 2010. Interactive contributions of genes and early experience to behavioural development: Sensitive periods and lateralized brain and behaviour. In *Handbook of developmental science, behavior, and genetics*, ed. G. Greenberg, C. Halpern, K. Hood, and R. Lerner, 400–433. Malden, MA: Wiley-Blackwell Publishing.

Rose, S. 1997. *Lifelines: Biology, freedom, determinism*. London: Allen Lane.

———. 2009. Should scientists study race and IQ? No: Science and society do not benefit. *Nature* 457:786–788.

Tooby, J., and L. Cosmides. 1990. The past explains the present: Emotional adaptations and the structure of ancestral environments. *Ethology and Sociobiology* 11:375–424.

Trivers, R. 1972. Parental investment and sexual selection. In *Sexual Selection and the Descent of Man*, ed. B. B. Campbell, 136–179. Chicago: Aldine.

Wehmer, F., R. Porter, and B. Scales. 1970. Prenatal and pregnancy stress in rats affects behaviour of their grandpups. *Nature* 227:622.

Weinstock, M. 2008. The long-term behavioural consequence of prenatal stress. *Neuroscience and Biobehavioral Reviews* 32:1073–1086.

Will, J. A., P. A. Self, and N. Datan. 1976. Maternal behavior and perceived sex of infant. *American Journal of Orthopsychiatry* 46:135–139.

Wilson, E. O. 1975. *Sociobiology: A new synthesis*. Cambridge MA: Harvard University Press.

Wood, J. L., D. Heitmiller, N. C. Andreasen, and P. Nopoulos. 2008. Morphology of the ventral frontal cortex: Relationship to femininity and social cognition. *Cerebral Cortex* 18:534–540. See also http://www.scientificamerican.com/article.cfm?id=girl-brain-boy-brain7print=true.

Wood, J. L., V. Murko, and P. Nopoulos. 2008. Ventral frontal cortex in children: Morphology, social cognition, and femininity/masculinity. *SCAN* 3:168–176.

Wood, W., and A. H. Eagly. 2002. A cross cultural analysis of the behavior of women and men: Implications for the origins of sex differences. *Psychological Bulletin* 128:699–727.

3

Looking for Difference?

Methodology Is in the Eye of the Beholder

BONNIE B. SPANIER AND JESSICA D. HOROWITZ

With increased legitimacy in studying homosexuality from a more nonjudgmental stance, much research over the past three decades asks: What are the biologically recognizable differences between gay people and straight people (sometimes adding a third, in-between category, bisexual people)? Asking the question that way is a choice that reflects a biological determinist (BD) view that different sexual orientations have a tangible "cause." The work of Dennis McFadden illustrates the quest for biological differences in sex and sexuality for the purpose of finding the cause of homosexuality.[1] In this chapter, we challenge McFadden's research on its methodological approach and its scientific methods to assess its scientific validity at this time. In doing so, we illustrate the power of methodology—the starting premises on which research is based—for research programs and, in particular, the privileging of group difference over individual variation in differences studies.[2] Here the concern for social justice merges with feminist critiques to improve science and will continue to do so as long as BD claims about groups differences are made.

The biological determinist claim that dichotomous differences in behaviors between specified groups, such as male-female, white-black, heterosexual-homosexual, are set and thus predetermined in some biological way can be critiqued solely on political justice grounds: differences from any chosen norm (male, white, heterosexual, able-bodied) have no bearing on human rights. Yet the view that societal morality and legal rights do not (or should not) follow from biology contrasts directly against E. O. Wilson's founding premise of the field of sociobiology in 1975: that evolution is (and should be) the basis for societal ethics.

In addition to political justice challenges, biological determinist claims about behavioral differences (including intelligence) have been challenged on the basis of "the science itself."[3] Since scientific knowledge is based on consensus, questioning the validity of research design and conclusions, as well as the premises underlying the studies, challenges the scientific validity of the claims. Therefore, we believe that inquiries into *how* questions are asked about human sexes and sexuality are crucial tools for eliminating societal constraints against full human development. At the same time, such analyses improve scientific objectivity by exposing assumptions that may bias or skew individual studies or entire research agendas.

Analyzing the validity of each new scientific claim about inherent biological differences between homosexuals and heterosexuals (or females and males) continues to be central to understanding how science contributes to cultural meanings of sex and sexuality. One can ask: (a) What biological differences constitute being a human female or a human male (or being black versus being white, or being gay versus being straight)? Or one can ask: (b) What biological variations constitute being a "normal" human evincing human behaviors related to sexuality? The difference reflects a fundamental methodological issue for social and natural sciences research on "sex."

Context is everything—in science as well as feminism. A scientist's research program is based on an explanatory framework and specific studies that build a particular worldview. Identifying and analyzing the assumptions or "givens" are key aspects of a critique of a study's methodology, while close scrutiny of the limitations of methods, data, and conclusions constitutes an analysis of basic scientific validity. We closely analyze the scientific work of one particular researcher, Dennis McFadden, who has studied male-female and heterosexual-homosexual differences in various auditory responses and ratio of finger-toe lengths. McFadden has conducted his research over the past twenty years in the contemporary socio-scientific culture, where biological determinants are ever more deeply assumed than in the 1970s and 1980s, when feminist and antiracist challenges were in greater evidence. Our analysis illustrates how methodological assumptions influence the interpretation of data, in some cases to the point of researchers ignoring inconsistencies in their own conclusions.

Most, but not all, feminists take historical context into account and point to biological determinism (BD) as one tool among many that justify beliefs against certain groups: women, Jewish people, blacks, gays, and lesbians. Some feminists in the fields of sociobiology (nonhuman animals) and evolutionary psychology (humans) have worked to eliminate sexism but do not challenge many BD assumptions (see, for example, Hrdy 1999). Politically, with the increased visibility of gays, BD claims are often used to promote gay rights.[4] Indeed, feminists assert that context and values shape research methodology. Our view is that feminist analysis of scientific claims, whether one agrees with the claims or not, is a powerful tool for social change.

Peer review of scientific papers for publication should address such limitations or flaws, but, as we shall see, acceptance of the more encompassing methodology often overlooks such problems. We suggest this occurs with McFadden's work on sexuality and BD since we identify serious limitations of the research, inconsistencies, and convoluted reasoning inadequate to support McFadden's working theory of sexuality.

Updating Challenges to Biological Determinist Claims

Investigations of biological bases for sex/sexuality differences in human behaviors continue despite scholars' delineation of a long history of poor science that does not stand up under scrutiny. Indeed, biologist Anne Fausto-Sterling imagined some validity to claims of sex differences in behavior and cognition because of that large body of scientific literature. She looked into every major research claim, such as greater brain lateralization in females than in males and better math ability in males, and published in 1985 an entire book, *Myths of Gender: Biological Theories about Women and Men,* that traced how each and every claim proved wrong or inadequate scientifically because of either contradictory evidence, insufficient evidence, or abandonment of theories. She, like most scientists, asserted we can distinguish between bad science and good science based on scientific content.

How far that holds is a judgment call. Critics of the science of biologically determined sex differences (as well as claims for race and sexuality) in behaviors and cognition have identified key conceptual errors that are a basic part of such research, errors not just in the methods chosen or the too-small sample size or inadequate attention to statistical significance, but in the more deeply embedded premises—the methodology—of the research.[5] Starting assumptions underpin how research questions are asked, how studies are constructed, which explanatory frameworks are acceptable and which are not, which data are highlighted and which data are excluded for being atypical, and how data are interpreted. Indeed, these critiques make visible the ways that biases in worldview enter the scientific method. The following key conceptual problems are found in the methodology of BD studies, and we will test McFadden's work with them.

Conceptual Errors in Biological Determinist Claims

I. REIFICATION. Making a complex, non-additive process into a measurable single thing, like IQ or sexuality, is highly questionable as scientifically valid. It is a fundamental mischaracterizing of a process and shapes how it can be studied.

II. CHOICE OF DEFINITION OF "DIFFERENCE." Differences between certain groupings (males, females, heterosexuals, homosexuals) are given priority over variations among all individuals in a population. How researchers ask the question is critical: How are women different from men in behaviors that can be related to biology? Or, How do individual humans vary in behaviors and biology across a population? This conceptual error of privileging "group difference" over "individual difference" across a population shapes how data is gathered and interpreted since overlaps between groups are ignored or minimized.[6]

More and more brain imaging studies claim to show male-female differences in brain activity. Anelis Kaiser and colleagues (2007) explored this common conceptual error as they reviewed research on language use and brain activity during varied language behaviors.[7] First and very importantly, they refuted the claim that such sex differences actually exist since larger studies as well as the most recent meta-analysis showed no sex differences; yet assertions continue to be made that such sex differences exist. Secondly, they showed through their own research with twenty-two males and twenty-two females that "differences" results are dependent on how the research question is asked, comparing male and female groups or comparing variation of individuals across the population of forty-four. The huge overlap or similarity between males and females is noted when individual variation is the method of comparison: "Seven females and nine males demonstrate bilateral patterns . . . while nine females and seven males showed left-lateralized activation. . . . [G]roup analysis enhanced the dissimilarities between the two groups, whereas the individual frequency analysis revealed similarities" (Kaiser et al. 2007, 196).

These researchers exposed the bias in their field toward sex differences rather than sex similarities and demonstrated that the bias existed in several ways. The prevailing definition of difference is a judgment affecting what is important enough to study, how studies are slanted to find group differences, and how some evidence is simply ignored. Thus, research on sex and/or sexuality differences starts with a methodological value judgment of group difference over individual variation. This overarching methodological choice is distinct from the next conceptual error.

III. QUESTIONABLE CATEGORIES. The categories used to compare groups are not distinct or mutually exclusive, so what is the significance of finding a "difference"? Males do things that are female-associated and vice versa, and cultures vary in what is "male" and what is "female," so "gender" varies greatly. Humans come in male and female biological sexes, but an estimated 2 percent are intersex, neither male nor female. The definition of "heterosexual" or "homosexual" is not fixed historically, such as viewing the whole being as a single sexual identity that does not change over the lifespan or defining sexuality by behaviors done with certain partners. What does "bisexual" mean in a society where most people experience same-sex behaviors sometime in their childhood or adulthood, and how does that affect being heterosexual?

In addition, definitions in research on sexuality are not consistent, with some studies including bisexuals in the homosexual category, and others not.

Some studies choose the very low estimate (1 to 4 percent) and others the larger percentage (10 percent or more) of homosexuals in a population, often depending on different definitions of "gayness."

IV. ASSUMING "UNIVERSAL" BEHAVIORS. This is a basic flaw from the start of sociobiology. For behaviors related to male-female "sex," only sperm donation or gestation and lactation are valid, and those do not apply to everyone in those categories. If you cannot match a behavior to a category, then you cannot say that *X* behavior predicts the person is *A* rather than *B*. All studies of group differences actually show an overlap between categories, as discussed in II above.

V. "NATURE AT BASE" AS AN INVALID CONCEPT. Everything exists within a social and physical context, even before birth. Studies of nonhuman animals are limited due to the rich cultural histories that shape humans' very plastic behaviors.[8] Anthropological claims about "primitive" peoples being closer to "nature at base" are questionable because so much cultural change has occurred since early humans evolved. Instead, the great variety of behaviors and social arrangements argue for human plasticity, such as our physiological capacity for language and music expression but in many different languages and forms.

VI. CORRELATION VERSUS CAUSATION. Most studies claiming to find a correlation are not accurate because of methods used or small size of sample, as Kaiser and colleagues suggest, or because the categories are invalid; so determining whether a correlation is actually valid is the first step before moving on to examine the validity of a cause-effect relationship. If the correlation is valid, that does not mean that the biology caused the effect because the biological change may be the result of the behavior. We now know that, unlike previous assumptions, brain structures change as a result of behaviors. In addition, the correlation may be due to some other confounding factor, and not the one studied.

VII. VAGUE "POTENTIAL" AS GENETIC BASIS FOR DETERMINISM. "Potential" and "predisposition" are vague terms that are questionable in the case of behaviors since what will happen is not knowable until the end of a life.

VIII. IRRELEVANCE OF MULTIPLE GENES EXPLANATION. The claim that multiple genes, rather than a single pair, can explain the overlap of behaviors and correlations is irrelevant in light of the other conceptual flaws that point to confounding factors and societal forces, which cannot be eliminated.

IX. ASSUMING "BIOLOGY" MEANS FIXED. Much of biology changes over time, so equating "unchanging" with "biology" is erroneous. Conversely, assuming that a characteristic that remains the same over time means it is fixed in one's biology is also erroneous.

X. HETEROSEXUAL CONCEPTUAL ERROR SPECIFIC TO SEXUALITY STUDIES. "Sexuality" definitions are often based on heterosexism, the assumption that a male is a male because he is attracted to a female and vice versa. Thus, a homosexual male is less masculine by definition when he chooses another male, as that is "female" behavior.

Some critics assert that these conceptual, methodological errors in human differences research are so substantial as to invalidate most, if not all, claims of behavioral or cognitive differences between social groups. Nonetheless, close scrutiny of individual studies and bodies of research remains an essential task for determining the limitations of the scientific work itself and for tracking possible improvements in research on biological determinism and differences. We examine McFadden's work in light of these conceptual errors in his methodology and then turn to his specific methods as well to assess the scientific validity of his claims.

McFadden's Research

Searching online news for an example of recent biological determinist claims, for a class by Bonnie Spanier, Jessica Horowitz noticed on a CNN Health Story Page that new evidence was about to become public in a prestigious science journal, claiming that sexuality is biologically determined. That news coverage led Horowitz to researcher and psychology professor Dennis McFadden, who for more than thirty years has been exploring auditory function as it relates to a myriad of factors. The chosen factors—"masculinization," hormone exposure, sexual orientation, health and behavior, auditory structure, heritability, status as a twin, sex differences, finger-toe length ratios—suggest his concern to account for sex and sexuality differences with biology or physiology. His faculty Web page at the University of Texas, Austin, indicates that his interest in auditory function has evolved from general heritability to sexual orientation and, most recently, the effects of androgen exposure on auditory function in humans and animals.[9] We have chosen to analyze Dr. McFadden's studies on auditory function as it relates to sexual orientation and gender (McFadden and Pasanen 1998) in the context of his research earlier in the 1990s and then to compare his methodology to later work, ending with a recent review (McFadden 2008) to assess any change in worldview in relation to his claims about BD and sexuality.

Sexuality Difference Rather Than Sex Difference in Ear Responses

In 1998 Dennis McFadden and Edward Pasanen published a study in which they measured click-evoked otoacoustic emissions (CEOAEs), a physiological response to an auditory stimulus (a click), which other researchers had found to be stronger in women than in men. By studying heterosexual, homosexual, and bisexual women and men, they sought to determine a link between the subjects' sexual orientation and that particular auditory function. The paper is noteworthy because it was published in the *Proceedings of the National Academy of Sciences*, a highly prestigious journal of mostly the physical and biological sciences.[10]

The rationale for their study, spelled out as expected in the first portion of the paper, established their methodology in the first sentence: "Evidence continues to accumulate about the biological concomitants of human sexuality" (McFadden and Pasanen 1998, 2709). Their argument supporting this assumption included early 1990s claims of differences in brain structures between heterosexual and homosexual males (LeVay 1991) as well as estimates of heritability of homosexuality from twin studies, plus a reference to Dean Hamer's claim (1993) of a linkage of homosexuality to the X-chromosome for men but not for women. Challenges to the validity of each of these claims had been raised well before this 1998 paper (Spanier 1995a, among many). The rationale for implicating OAEs (otoacoustic emissions) with sexuality included the constancy of individual OAEs from birth, heritability from twin studies, and the claim of male-female differences. Thus, something that remains particular to an individual's physiology is seen as a biological given, and the claim that ear emissions differ from men to women made them potential markers of sexual orientation, first by the implication that sexual orientation is a biological given also fixed at birth and second by the inference that sex/gender difference is related to sexual orientation difference.

The most compelling foundation for their methodology was their own research showing that in opposite-sex dizygotic (OSDZ, fraternal) twins, the females have two kinds of OAEs more similar to males than to females (McFadden 1993; McFadden et al. 1996). The authors had proposed earlier that the androgens from the male twin had affected the female twin through the amniotic fluid they shared, an effect known to occur in twin calves (producing a sterile female "freemartin") and other mammals. (We found that this foundational research was not so clear-cut, as discussed below in the section "Limitations, Contradictions, and More.")

Their 1998 results indicated a variation in CEOAE strength ranging from strongest in heterosexual women and then weaker in homosexual women, bisexual women, bisexual men, homosexual men, and heterosexual men. The researchers concluded that, while the CEOAEs were similar for both homosexual and bisexual women, they were weaker than the CEOAEs for heterosexual women. The researchers asserted that these ear emission levels were related to sexuality in females. They found no such difference among males of different sexual orientations.

The authors devoted much space in the paper and the abstract to the claim that their results support a theory that the nonheterosexual females' lower responses indicated they were "masculinized" and that this effect was due to "exposure to high levels of androgens prenatally" (McFadden and Pasanen 1998, 2709). While they tried to account for their finding that males of differing sexualities had the same response and to also address other inconsistencies, they did not waiver from their main claim, which is that nonheterosexual females are masculinized because of fetal androgens. Most notable is the authors' assertion that "the present CEOAE data can be interpreted as evidence that prenatal exposure to higher-than-normal levels of androgens in homosexual and bisexual females produced a partial masculinization of both their peripheral auditory systems and some brain structures involved with sexual orientation" (2712).

The study and the complexity of the measurements are detailed well in the paper, and so provide specifics for critiquing methods as well as methodological assumptions.

Methods

The researchers recruited 291 subjects through gay publications, university newspapers and bulletin boards, and local gay groups. Some (14 percent) were eliminated immediately with a hearing test, while 13 more were found to have CEOAEs that did not meet certain criteria, including "peculiarities in their CEOAE waveforms" and inconsistencies across different decibel levels used to evoke the responses (McFadden and Pasanen 1998, 2710). A total of 237 subjects were assigned—from their self-identification and other answers about sexual experiences and sexual fantasies (based on Kinsey questionnaires)—to one of these categories: heterosexual female, bisexual female, homosexual female, heterosexual male, bisexual male, homosexual male. The ear study was done by inserting into the ear canal a tiny microphone (the receiver) with a tiny earphone that administers small click sounds; the microphone detects the physiologically evoked responses that are measured as a series of echo-like waveforms. The measurement of the CEOAE waveforms is highly complex and variable (suggested by the exclusion of 13 normal-hearing individuals as well as many measurements), producing a "mean CEOAE" value that was repeated until there were 250 responses "satisfying the collection criteria."[11] It was known that intense noise, certain drugs, and other factors affect the response, and that right and left ears produce distinct responses (so each ear was tested, as the figures show). Four different decibel levels of clicks were used, and each was reported in pairs in figures 3.1 and 3.2. The mean CEOAE data were calculated for the right and left ears of each individual and for each decibel level.

Then the final step to produce data involved selecting a particular portion of each averaged waveform and converting it to a sound-pressure level (SPL)

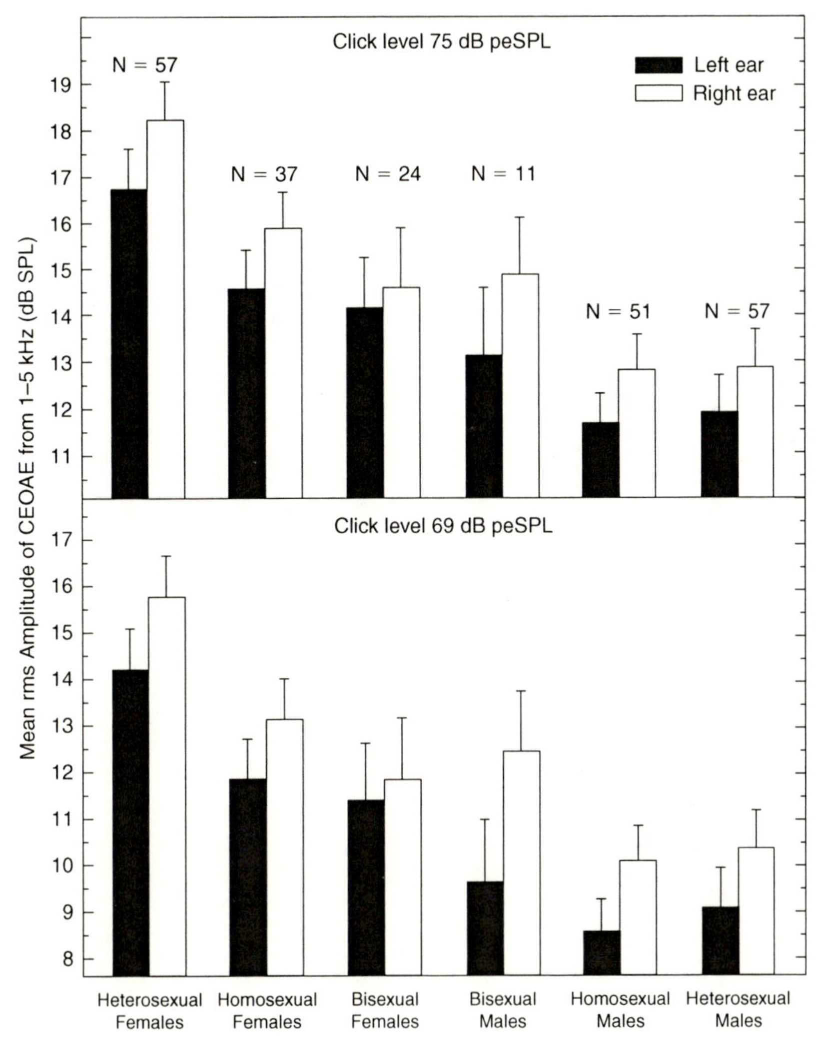

FIGURE 3.1 The rms amplitude in the averaged CEOAE waveforms for the two highest click levels tested (75 and 69 dB peSPL), averaged across all subjects in each group. The analysis window was from 6.0 to 27.3 ms following the presentation of the click. Responses to 250 clicks were collected for each click level. The error bars indicate one standard error.

Source: D. McFadden and E. G. Pasanen, Comparison of Auditory Systems of Heterosexuals and Homosexuals: Click-Evoked Otoacoustic Emissions. *Proceedings of the National Academy of Sciences U.S.A.* 95 (1998): 2711.

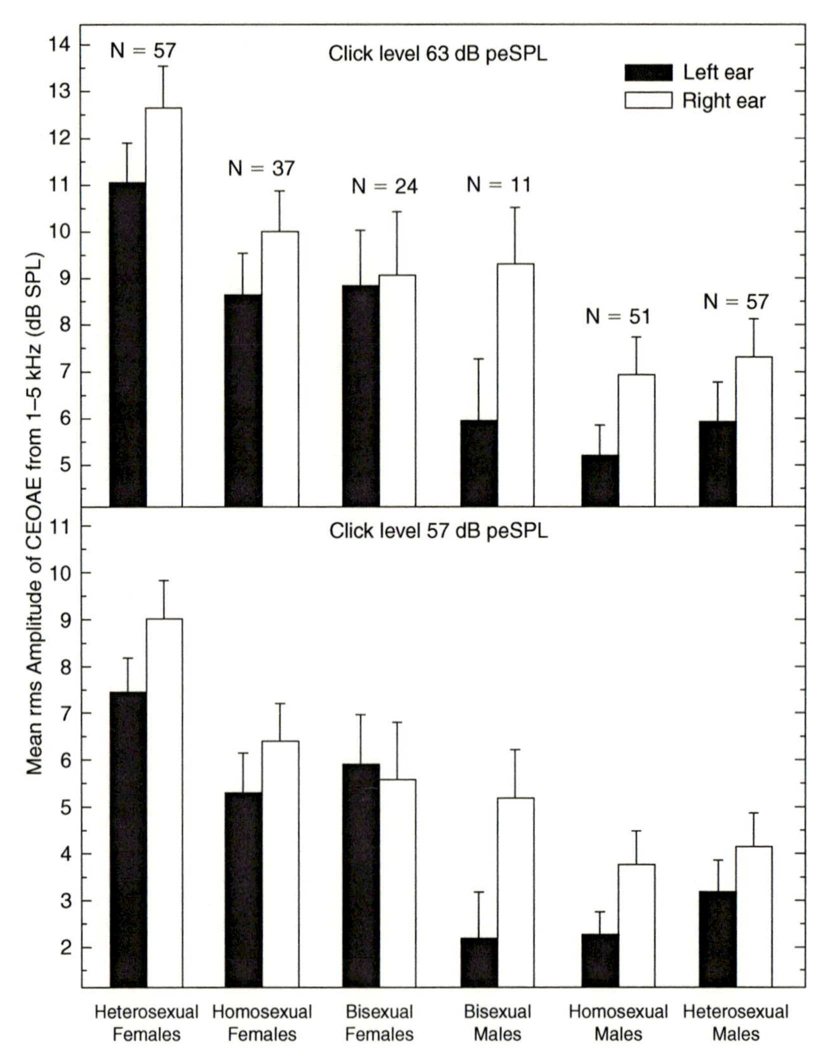

FIGURE 3.2 The rms amplitude in the averaged CEOAE waveforms for the two lowest click levels tested (63 and 57 dB peSPL), averaged across all subjects in each group. Otherwise, details are the same as for fig. 3.1.

Source: D. McFadden and E. G. Pasanen, Comparison of Auditory Systems of Heterosexuals and Homosexuals: Click-Evoked Otoacoustic Emissions. *Proceedings of the National Academy of Sciences U.S.A.* 95 (1998): 2711. © 1998 National Academy of Sciences, U.S.A.

value, and those SPL values were averaged across the group "to summarize the strength of response in each group" and reported for each sexual orientation/ sex group (McFadden and Pasanen 1998, 2710). Thus, mean CEOAEs and SPL values are reported only by group and not by individual.

Results

Figure 3.1 shows the aggregate data (for two decibel levels) as the mean for each group's right and left ears, with the standard deviation indicated. (Figure 3.2 shows the results for the other two decibel levels.) Note the *N* above each bar, indicating the number of individuals in each category. The variation in CEOAE strength is presented as ranging from strongest to weakest: heterosexual women, homosexual women, bisexual women, bisexual men, homosexual men, and heterosexual men, with 11 to 57 subjects in the groups.[12] McFadden and Pasanen stated that the men did not show the sexuality difference in CEOAEs that the women showed, a male-female difference in sexuality causation that they must address as inconsistent with their theory (see below). They used the concept "masculinized" to explain the nonheterosexual females' lower CEOAEs and then worked to retain the theory to account for the male results.

Prenatal Androgen Theory

In discussing the research findings, McFadden and Pasanen stated that "the present CEOAE data can be interpreted as evidence that prenatal exposure to higher-than-normal levels of androgens in homosexual and bisexual females produced a partial masculinization of both their peripheral auditory systems and some brain structures involved with sexual orientation" (1998, 2712). Aware of the two pieces of refuting evidence to this theory, the researchers immediately clarified their point: "It is intuitive that different brain sites would be masculinized at slightly different times in prenatal development, meaning that androgen levels must be adequately high in each time period for all sites to be masculinized fully" (2712) (see discussion of table 3.1). In response to the issue of gay men having a nonmasculinized sexual orientation, the researchers stated: "Male homosexuality still might be attributable to a deficiency in prenatal androgen exposure at some critical brain site" (2712). These inconsistencies are discussed below in the section "Limitations, Contradictions, and More."

Critique of Methodology (Conceptual Errors) and Methods of 1998 CEOAE Study

McFadden and Pasanen's exploration into the link between CEOAEs and sexual orientation exemplifies the perils of allowing value-laden methodological assumptions to infiltrate what should be a value-neutral inquiry. Using the list of ten conceptual flaws of BD studies, this research demonstrates errors of reification (I), choice of group definition of "difference" over individual variation (II), questionable categories (III, with IV and V embedded), correlation versus causation (VI), assumption of biology as fixed and unchanging (IX), and, finally, sexuality definitions based on sexism/heterosexism (X) that assume sexual orientation or preference is tied (biologically) to sex/gender definitions (male is not male enough if he chooses another male).

The authors work within an explanatory framework of sexuality as biologically determined (or influenced). They reify sexuality (I) by assuming it is a single attribute of a person, rather than a set of behaviors that can be fluid and dynamic throughout one's life. Their language choice of "sexual orientation" rather than "sexual preference" reflects their assumption about sexual behavior as a fixed identity of the human being, rather than a set of behaviors with a type of partner (same or other sex) that can and often does vary in a lifetime. Notably, study subjects were in their twenties, so the assumption is underlined that one's sexuality is fixed and will not change. Arguments about what sexuality *is* (identity, orientation, preference, etc.) have been common in and out of lesbian and gay, as well as dominant, communities for a long time, revived in the public eye after Simon LeVay's 1991 *Science* paper. Throughout, McFadden never raises this possibility that sexuality is not a trait fixed for life. He does recognize individuals who are neither entirely heterosexual nor homosexual, but creating a "bisexuals" category for males and for females—and then merging it with homosexuals—does not change his methodology. And he studied transsexuals (McFadden et al. 1998) for estrogen effects on ear emissions without mentioning how transsexuals might complicate binary and nonbinary sex and sexuality systems.

With regard to his clear choice of group difference over individual variation (II), this paper is striking in the absence of any individual data. Nor is the data ever presented in a way that would address the extent of overlap between groups. To an extent, this absence of individual or group overlap data may be influenced by the complexities of the CEOAE measurements (at least that is intimated in the methods section), but whatever the reason, the absence of individual data serves to obscure whether actual group differences were found. Indeed, the standard deviations on the bar graphs indicate there is no apparent statistically significant difference between most of the groups. In fact, if one lays out the data and tries to apply the authors' theory of androgens' masculinizing effects prenatally, we generate table 3.1. What is striking here is that heterosexual females emerge as the only "different" group, while the others—bisexual females, homosexual females, bisexual males, homosexual males, *and* heterosexual males—are similar to each other in "masculinized" CEOAEs. But the authors do not recognize that data result at all. It would not fit with their chosen groupings based on sex and then broken down into sexual orientation.

TABLE 3.1

Contradictions produced by heterosexist methodology

Sexual orientation and sex	*Masculinized CEOAE*	*Masculinized sexual orientation*
Heterosexual female		
Bisexual female	X	X
Homosexual female	X	X
Heterosexual male	X	X
Bisexual male	X	X
Homosexual male	X	

Note: X indicates a "masculinized" result.

McFadden's Conceptual Errors Common to Biological Determinist Claims

DETERMINING AND CATEGORIZING SEXUAL ORIENTATION (III). The problem of questionable and shifting categories is major throughout McFadden's work.[13] Putting aside the cultural construction of maleness and femaleness to focus on categories of sexuality difference, and leaving until later the conceptual errors of sexism/heterosexism (X), we can appreciate the magnitude of the problem in a series of studies where researchers are trying to "explain" different sexual orientations with the theory that prenatal hormones made them "that way."

The authors' methodology shows no sign of acquaintance with feminist theorizing about sexual orientation, such as Adrienne Rich's (1980) now-classic insight that compulsory heterosexuality belies free choice and immediately challenges definitions of "sexual orientation." Rich's concomitant concept of the lesbian continuum further complicates the question of who is a lesbian, such as women who considered themselves totally heterosexual until they actively chose a lesbian lifestyle in the 1970s women's liberation movement, much less chaste nuns devoted to one another. Taking Rich and her feminist theoretical legacy (with its disagreements and appreciation of the power of internalized homophobia and heterosexism) into account would have also reminded the researchers that coming out as a nonheterosexual remains problematic (in different ways for males and females), so it is reasonable to suggest that some of the heterosexual women and men in the study were not heterosexual, but chose consciously not to reveal their identity or behaviors as such. Since we cannot assume that all the self-identified heterosexuals were actually heterosexual, those female and male group categories are suspect simply at the level of accuracy in methods.

Another aspect of our challenge occurs at the level of methods. To determine sexual orientation, the researchers created a questionnaire based on self-identification and Kinsey items on sexual experience and sexual fantasy. We are skeptical that any reliable measure exists in the first place, since sexuality is culturally constructed and highly fluid.

CORRELATION VERSUS CAUSE (VI). The first step in a critique is to determine *if* the data show a statistically significant correlation. Where are the valid correlations in this paper? The data show a correlation between heterosexual females and the strongest CEOAEs, while lower responses correlate with *all* the other groups without being able to make a clear distinction among them. Perhaps larger size groups would provide more information, but, in this study, with the overlap from standard deviations as well as the great variation when comparing left versus right ear results at difference decibels, the data do not show significant group differences in CEOAEs. Simply put, the fact that heterosexual women have CEOAEs that are significantly stronger than homosexual and bisexual women suggests that CEOAE strength is *not* based on sex—if it were, then all women would have levels that were more or less the same and significantly different from all males. The data do not support consistent male-female difference in CEOAE strength.

Most importantly with regard to claims about what is causing the stronger CEOAEs in female heterosexuals, we question the validity of the authors' proposal that this study of differences in ear emission strength supports the theory of prenatal exposure to androgens as the cause of sexual orientation (but only for females). Even setting aside our critique of the study's methodology and methods, the data are not strong or consistent with their theory, as further delving into limitations, contradictions, and inconsistencies indicates below.

ASSUMPTION THAT WHAT IS FIXED AND UNCHANGING IS "BIOLOGICAL" (IX). In addition to the premise that sexual orientation is a fixed trait that has a cause at the level of hormone action on the fetal brain, setting identity and behaviors for life, the authors search for physical characteristics that remain unchanged throughout a lifespan. That is their rationale for constructing this study of differences among differing sexual orientations around CEOAEs, which they claim remain the same from birth on. We can ask: Just how fixed are each person's CEOAEs (and other auditory measures that come up in later studies), given the known modulating effects of certain drugs, intense noise, congestion, as well as unknown modulators? If so many subjects (about one in six) were eliminated from the original group because of inadequate hearing and (unexplained) peculiarities in responses, how reliable is it to claim that such measurements are fixed traits? Further, with so many measurements excluded for weakness and other reasons as part of the study, how reliable are the measurements at all?

The fundamental challenge to the assumption of fixity equaling embedded biological trait pertains to every aspect of these BD studies.

SEXUALITY BASED ON SEXIST/HETEROSEXIST ASSUMPTIONS (X). Spanier has argued elsewhere that good politics do not justify bad science, or, conversely, that bad scientific evidence should be ineligible to support good politics (Spanier 1995a). Thus, just because we identify sexist and heterosexist beliefs in the study's methodology does not make it invalid scientifically; it may just make it distasteful politically. But studies run into scientific validity problems with the view that there are two sexes that are in some fundamental way inherently different from each other and that one is naturally "male" or "female" by virtue of being attracted to the "other" sex.

The intertwined and also confounding nature of sexism and heterosexism is in evidence. As a basis for their work, McFadden and Pasanen cited the accepted view that there are differences in otoacoustic emissions based on sex. But their 1998 study contradicted the earlier male-female difference claim, unless you take only heterosexual males and females as valid males and females! Despite the *lack* of uniform correlation of CEOAE strength with sex when the groups are divided according to sexual orientation, the researchers continue to discuss the gender binary as though they had not just revealed results that cast doubt on it. The authors' dependence on sex difference, based in androgen difference, creates internal contradictions between their theory and their data, which the authors largely ignore.[14] An adherence to the gender binary is not unusual in BD science research, but what makes this case especially interesting is the ways in which the discussion of results is further influenced by both the sexist idea of male and female difference and heterosexist assumptions that sexuality is dependent on sex designation. (We find the term "hetero/sexism" useful to indicate the inseparable nature of sexist and heterosexist beliefs.)

Hetero/sexism is in evidence as well in the authors' gender-biased language that obscures accuracy and promotes their unproven theory.[15] Despite female-associated CEOAEs being *stronger* than those of males, they avoid the more accurate descriptive terms, "weak" and "strong," and replace them with the term "masculinized" to characterize nonheterosexual females as less female-like and more male-like. In so doing, the researchers' language choices superimpose onto the data results their view that CEOAE magnitude is based on sex.

CONTRADICTIONS THEY ACKNOWLEDGE AND RESOLVE. The researchers recognize two contradictions (or exceptions or inconsistencies) that arise from the evidence in relation to their theoretical claim of the apparent masculinization found in this study. Interestingly, after they acknowledge these two problems with the evidence, their solution is to offer another theoretical refinement to the theory, adding the factors of timing and magnitude of exposure to prenatal androgens.

The first contradiction stems from the CEOAE results regarding the homosexual men: no statistically significant difference between the CEOAEs of the homosexual men and those of the heterosexual men. So the homosexual men in the study have the masculinized (weak) CEOAEs, but they *do not* have a masculinized sexual orientation.

The second related contradiction comes from McFadden's 1993 and 1996 studies of twins, claiming to find that the female twins had OAE strength similar to their male co-twins, thus masculinized females in their OAEs (McFadden 1993; McFadden et al. 1996). Logically, with the OSDZ twin study as evidence for prenatal androgen exposure, McFadden and Pasanen's (1998) assertion of evidence for nonheterosexual females having masculinized CEOAEs means that the OSDZ females should show nonheterosexuality, but they do not.

How do the authors address these inconsistencies that raise doubts about their theory? First, the authors suggest the theory could simply be wrong, but then, without missing a beat, they assert that "it is intuitive" that *timing* for androgen exposure at different sites in the brain may be the explanation. This easy throw-away line, which suggests that if you don't see it, you are missing a major point, dismisses the problem that the data and interpretation are not consistent with other studies in supporting the prenatal androgen theory of sexuality and ear emissions.

Limitations, Contradictions, and More

UNCONVINCING OSDZ STUDIES. Much more significant is a problem one finds when examining McFadden's original 1993 and 1996 papers on the eighteen OSDZ twin pairs, the basis for the theory of fetal androgens affecting sexuality and ear emissions. *No* such correlation was found for CEOAEs in the OSDZ females (McFadden et al. 1996), while a statistically significant correlation was found for spontaneous otoacoustic emissions (SOAEs) (which are of different origin from CEOAEs) *only if one twin was excluded* for having such a high result that it was almost eight standard deviations from the rest (McFadden 1993). *With that twin in the calculation, no masculinizing effect was seen.* This result suggests the tenuous nature of claims from that 1993 paper, which was published in the prestigious *Proceedings of the National Academy of Sciences,* like the 1998 paper. With no effect in evidence for CEOAEs (which were the subject of the 1998 study) and with the questionable effect on SOAEs (dependent on excluding one single subject), the premise that the twin studies are the foundation for the fetal androgen theory of sexuality is highly suspect.

DIFFERENCE, BUT NOT A LINEAR SPECTRUM. The presentation of data in their two figures (figs. 3.1 and 3.2) characterizes the results as a spectrum, and this is actually misleading because the differences among those groups other than heterosexual women are not significant, especially if you look separately at each set of results for each ear and at the four decibel levels (that is, you could actually change their response order). Thus, the only difference they find in the 1998 paper is between heterosexual women and everyone else. Indeed, McFadden's continued work on BD and sexuality changes the categories to heterosexual and nonheterosexual, grouping bisexuals together with homosexuals, yet the authors retain the male-female separations along with sexuality.

ADHERENCE TO THE THEORY OF FETAL ANDROGENS AND SEXUALITY. Male and female sex hormones provide a cornerstone of BD explanations of sex differences and then sexuality differences. Unfortunately for the theory, *no* evidence supports it, despite decades of searching for hormone imbalances in gay men and lesbians.[16] Theoretical *in utero* effects of hormones to "set" the fetal brain on a male or female course of development and subsequent behaviors became the next cornerstone. The fetal androgen theory of sex and sexuality is based on a long-known observation of the freemartin calf, a masculinized female OSDZ calf (as well as studies of lab rodents injected with hormones and possibly showing changes in male-typical and female-typical body positioning). The effect does not seem to occur in many mammals with multiple fetuses. If such an effect occurred in humans, it would be obvious and well known by now, given general observations of human twins and more scientific studies. One can only marvel at the fortitude of some researchers like McFadden and the funding agencies that continue to validate the question: What makes them (or us) gay? rather than Adrienne Rich's insightful question: What makes anyone heterosexual?

INCONSISTENCIES SIGNALING ALTERNATIVE EXPLANATIONS. In fact, if dichotomous scientific explanations weren't so common, a reader might look at the results and find that the anomalous group (the group that showed a marked "difference"—no value judgments) is the heterosexual women (because they have stronger CEOAEs than the other five groups included in the study). But because of heterosexism, a member of the heterosexual group is not relegated to "other," as the authors do not address this issue. Yet it is just this kind of result that should make an open-minded scientist ask: Are there better explanations for the body of results? That the authors do not waiver from their commitment to the fetal androgen theory suggests their rigidity.

Still Searching for BD and Sexuality Evidence

Since 1998, McFadden has continued his search, measuring the ratios of fingers and toes, ear emissions, and other responses to auditory stimuli in transsexuals (but less than a handful) and others possibly affected by hormones (such as women taking birth control pills). His work remains characterized by a dogged adherence to the fetal androgen theory of biologically determined sex and sexuality differences, complex manipulation of ear emission measurements, and selective exclusion of subjects or measurements. The worst example is McFadden and colleagues' 1998 study, where the subjects consisted of two transsexual males undergoing transition to female with estrogen treatments. One male showed no change in SOAEs and so was essentially excluded, with the rationale that he was not assessed before starting estrogen treatments. The other transsexual showed a shift to female-type emissions. What does it mean for the state of scientific research that this single case study of ear emissions of a transsexual in transition—an individual with "considerable hearing loss in both ears" (McFadden et al. 1998, 1557)—got published in a peer-reviewed journal? In addition, the authors changed their categories of difference (deleting bisexual and only comparing heterosexual to nonheterosexual) while highlighting only the results that supported sex difference in addition to sexuality difference.[17] The same methodological issues apply throughout as we detailed for the 1998 paper by McFadden and Pasanen. In 2008 McFadden published an overview of his work and related evidence, giving us an excellent opportunity to ask whether he has modified his claims.

McFadden's View in 2008: "What Do Sex, Twins, Spotted Hyenas, ADHD, and Sexual Orientation Have in Common?"

In this recent review McFadden reasserts his longstanding proposal that there is a "relationship between androgen exposure and OAE strength" (2008, 309) by providing an overview of his research on "special populations." He reiterates, too, his claim of a correlation between OAEs and fetal androgens. Then he makes a special plea for parsimony (318). We focus here on his work and claims about humans, sexuality, and ear emissions.[18] In the humorous and self-deprecating tone of the review paper, he comments that his measures to probe human development and interaction with the world of stimuli "are unquestionably imperfect, but we now have measures that seem to be sensitive to the degree of exposure to androgens a developing fetus receives." Then he adds, "Some readers surely will find it bizarre that those measures come from the auditory system" (309).

Most striking is McFadden's admission that his primary theory relies on evidence that is "largely circumstantial" (2008, 309).[19] He espouses some limitations of his work, but not the ones we have delineated here, particularly the tenuous foundation of OSDZ females. His flimsy argument appears more and more to be a house of cards built with insubstantial and contradictory evidence. He does credit himself with rejecting the direct method of investigation, which would involve manipulating prenatal androgens in humans, which explains why he settled on studying "people who were exposed to atypical hormone levels during prenatal development because of some malfunction of the ordinary development process" (312).

A new piece of logic emerges when McFadden reasons that since all fetuses are female (by default) then "the default choice for sex partner is male" (2008, 316), a statement based entirely on heterosexism. He goes on to say that androgens, in addition to changing the default sex from female to male, also somehow affect the change to the nondefault option in sex partner. McFadden uses this totally hypothetical logic to conclude that "prenatal androgen exposure somehow produces a masculinization of the brain region(s) responsible for choice of sex partner in humans" (316). Here he is simply explicating his reasoning for his theory that "these physiological concomitants suggest that some fundamental biological processes contribute to the existence of homosexuality" (317).

McFadden admits to major uncertainties several times, such as, "No one yet knows for certain when in life the changes in the nonheterosexual brain occur, nor how or where," and to "having been overly simplistic" (2008, 317) in his discussion of masculinization and the many ways that ear emission changes can occur: oversensitive receptors, globally high androgen levels, exposure at different periods in development, exposure to different regions of the brain, and aromatization. Despite that list of uncertainties, he refers once again to the nonheterosexual female and states that "whatever the exact mechanism, however, the end result would be a masculinization of the affected brain regions (plus the auditory system), and it is that masculinization event that is crucial to this explanation of the atypical choice in sex partner in nonheterosexual women" (318). He even reminds the reader that there are alternative explanations for the variation in OAEs (exposure to loud sounds, hearing loss, ototoxic drugs).

Finally, McFadden comes back to a rationale that he often raises in many of his studies to support his theory: parsimony. After acknowledging the possibility that there may, in fact, be no common underlying cause for all the different observations, McFadden reminds the reader that until the answer is found "science dictates that one tries to find the simplest possible explanation for as many facts as possible, and the prenatal-androgen-exposure explanation appears to do the job" (2008, 318). He tells us to simply accept the parsimony assertion, and then claims without any justification that his theory is the most parsimonious. But, skeptics may ask, is that true? To us, the fluidity of sexuality with the social construction of meanings of maleness and femaleness is a far more valid parsimonious explanation for variations in sexuality across populations and across cultures and across lifetimes.

McFadden boldly states that his "explanation now appears to be solid enough for all of us to become serious about devising tests to disprove it" (2008, 318). McFadden has done a masterly job of seeming to admit limitations in his research and theory, while he ends up solidifying the correctness of his theory to the point that it is up to critics to *disprove* it! Far from evincing openness to the limitations, contradictions, and convoluted reasoning of his claims, he reveals the rigidity of his belief in his worldview. His methodology has not budged an inch.

Conclusion

McFadden's body of work is not his alone. His work (and that of his co-authors) has been long supported by grants from one of the U.S. government's National Institutes of Health, the National Institute of Deafness and Other Communication Disorders. In addition, every published paper represents approval from the scientists who peer-reviewed the manuscript and the editors who made the final decision to publish it. We suggest from our critique of McFadden's work that biological determinist (BD) research on sex and sexuality group differences is alive and well in the scientific community; and the search for the biological "causes" of nonheterosexuality (with heterosexuality as the norm that needs no explanation) continues on its convoluted path, oblivious to signposts of insights from feminist theorizing and supported by our tax dollars, heterosexual and nonheterosexual alike.

ACKNOWLEDGMENTS

We would like to thank the Community Health Library of Munson Community Health Center in Traverse City, Michigan, for assistance in locating primary articles and K. Smilla Ebeling for bringing Anelis Kaiser and colleagues (2007) to our attention.

ACRONYMS

ADHD: attention deficit hyperactivity disorder
AEP: auditory-evoked potentials
BD: biological determinist
CEOAE: click-evoked otoacoustic emissions
fMRI: functional magnetic resonance imaging
OAE: otoacoustic emissions
OSDZ: opposite-sex dizygotic
SOAE: spontaneous otoacoustic emissions
SPL: sound-pressure level

NOTES

1. In this chapter, we often refer only to McFadden as he is the major researcher on the several papers we critique. Note that McFadden's body of work discussed here includes single-authored and co-authored papers, but McFadden is first author on all of them (see references).

2. We base our approach on feminist theorists who have highlighted the power of epistemologies (i.e., who can shape the research questions and priorities and what constitutes legitimate knowledge) and methodologies (i.e., what assumptions are embedded in the research and how the research question shapes legitimate answers) to create partial and impartial knowledges (e.g., Harding 1987).

3. We acknowledge the spectrum of views about the validity of scientific information, from an absolute belief system that privileges scientific information as the most dependable to the other extreme of science as a product of only cultural beliefs. We take a middle ground to engage with empirical science as it is shaped by deeply held beliefs about human nature.

4. This is to the point that some antigay religious groups adopted the view that, since homosexuals are born "that way," they cannot be blamed but are still expected not to act on their impulses because homosexual behaviors are wrong and unnatural.

5. See the work of Anne Fausto-Sterling, Ruth Bleier, Richard Lewontin, and Stephen Jay Gould.

6. A prime example is comparing men's and women's height. Height variation *within* the group "men" or *within* the group "women" may be eighteen to twenty-four inches (even excluding small people, who should be included), while the average difference in height *between* men and women is two to four inches. Difference within each group is far greater than the average differences between the two groups. So it is misleading to focus on the view that men are taller than women, since the overlap between the two groups is so great as to render the claim inaccurate.

7. This area of differences research in "behavior" does not involve performance differences, but differences in regional brain activity (detected by functional magnetic resonance imaging [fMRI]) while performing the same functions. Thus, despite the fact that the claim of women's brains being more lateralized than men's was shown to be spurious (because studies were inconsistent), new subtleties keep emerging in the search for male-female differences as in lateralization of brain *activity*.

8. "Plastic" and "plasticity," used in biology and neurobiology in particular, indicate fluidity, the opposite of rigidity.

9. See http://www.psy.utexas.edu/psy/faculty/Mcfadden/mcfadden.html (accessed November 14, 2010).

10. In addition to the usual peer review, submissions to *Proceedings of the National Academy of Sciences* must be sponsored by a member of the U.S. National Academy of Sciences.

11. This is clearly a complex process using pre-amplifier, amplifier, and filter, and then an analog-to-digital converter to generate a digitized waveform. Clicks are done at ten per second with random intervals, and the echo-like responses are collected and summed and also spaced to avoid interference: "The scheduled click was presented only if that rms value was below that subject's noise criterion. A similar procedure was used to reject individual responses to the clicks" (McFadden and Pasanen 1998, 2710).

    Right and left ears differ significantly in response, as figures 3.1 and 3.2 show, so for each ear a click level is calibrated based on all these factors. Thus, much internal selection of data took place as a normal part of the process, and data on the variation in the basic process is not apparent. And many factors modulate CEOAEs temporarily or permanently, such as drugs and intense sounds.

12. Number of subjects in each group is as follows: heterosexual women, 57; homosexual women, 37; bisexual women, 24; bisexual men, 11; homosexual men, 51; and heterosexual men, 57.
13. Claims of universal traits for different groups and some biologically basic identity causing those traits or behaviors are embedded in the category problems of BD.
14. See note 17 for a similar problem.
15. They construct maleness as active, a stereotypical stance (Spanier 1995b). McFadden and Pasanen wrote: "Male fetuses naturally produce high levels of androgens at specific points in prenatal development and thus *are responsible for diminishing their own CEOAEs*" (1998, 2709, emphasis added).
16. The old and simplistic nature-nurture argument about sex differences and sexuality differences was all about nature and hormones in the 1950s and then shifted to nurture by the 1970s, with theories such as the cold father and smothering mother causing male homosexuality, for example. Dr. John Money, head of Johns Hopkins University's Gender Identity Clinic, pushed the nurture side to answer what makes a girl a girl (just raise the child as a girl, as long as there is no penis) with his case study of a biological boy, an identical twin, raised as a girl after his penis was damaged. Money reported for years the great success of this unfortunate accident, with stories of a happy and normal girl. In 2000 the full story was heard when the truth about the child (not knowing what had happened to "her") as horribly unhappy and suicidal was revealed; she had chosen in her teens to live as a boy, rejecting the surgery her parents had been pushing at Dr. Money's insistence. The man married a woman, had reconstructive surgery to create a penis, and lives a "normal" male life, with a notably feminist understanding of constraints on little girls and women. His startling story, which finally discredited Dr. Money, is now interpreted as the death knell for nurture in sex identity (Colapinto 2000). But it just isn't that simple. Nonetheless, throughout the last fifty years the predominant explanation for gender identity has been the effect of male sex hormones shaping the fetal brain, "masculinizing" it, and producing whatever behaviors society calls "male." The jump from lab rodents and monkeys to humans is huge when one considers the construction of maleness and femaleness in societies and the co-construction of sexuality with sex/gender—the assumption that a true female is attracted to males only, and vice versa.
17. McFadden and Champlin (2000) studied sexual orientation and brain (not ear) responses to sounds, called auditory-evoked potentials (AEPs), quite different from OAEs but involving even more complex types of measurements. They started testing the same six categories of sexuality as in 1998 with CEOAEs, but ended up with only four groups (heterosexual and nonheterosexual for both males and females) because of the small number of male bisexuals and also the similarity of responses of the homosexual and bisexual females. This demonstrates the privileging of difference over similarity, a major problem in differences methodology.

    Only a subset of five measurements (out of the nineteen measured) were used, chosen for their demonstration of the differences the researchers were looking for, differences between heterosexuals and nonheterosexuals. Thus they excluded data showing no difference. All five measurements were used for the males, but only four for the females. Of these, some showed a basic sex difference along with a sexuality difference, but others did not show the sex difference. Referring to those cases where there was a difference based on sexual orientation but not on sex, McFadden and Champlin stated that "those measures provide little insight into the mechanism underlying the differences between heterosexuals and nonheterosexuals" (2000, 95), again skewing their data by selective exclusion that limits the possible meaning of available data. In their theoretical view, the lack of a basic sex difference in their measures indicates a lack of impact by androgens; and, therefore, there is no relevance to the theory that sexual orientation (at least nonheterosexual sexual orientation) is affected by androgens. This is another remarkable example of selective data use in order to support a theory; good science requires taking all data into account to judge among alternative explanations.

This AEP study is very complex and involves results labeled "hypermasculinization." See Horowitz's forthcoming master's thesis in women's studies at the University at Albany.

18. The other two pieces of evidence come from animal studies. Observations of female spotted hyenas (an animal that shows masculinized traits such as a larger-than-male body size, dominant behavior, and an enlarged clitoris that "is traversed by a single canal that serves for mating, birth and urination") (McFadden 2008, 313) revealed that they have slightly weakened OAEs. A study on rhesus monkeys showed that, when exposed to a higher level of prenatal hormones late in gestation (via purposeful manipulation by the scientist), the monkey's OAEs were masculinized.

19. The circumstantial evidence, other than the OSDZ plus the 1998 sexuality paper, is McFadden's research on twin boys with ADHD (attention deficit disorder), which revealed "hypermasculinized" OAEs. McFadden has proposed that the boys "may have been exposed to higher-than-normal levels of androgens at some point in early development, perhaps during prenatal development" (McFadden 2008, 315), again, a completely hypothetical assertion!

REFERENCES

Colapinto, J. 2000. As nature made him: The boy who was raised as a girl. New York: HarperCollins.

Fausto-Sterling, A. 1985. Myths of gender: Biological theories about women and men. New York: Basic Books.

Hamer, D. H., S. Hu, V. L. Magnuson, N. Hu, and A.M.L. Pattatucci. 1993. A linkage between DNA markers on the X chromosome and male sexual orientation. *Science* 261: 321–327.

Harding, S., ed. 1987. *Feminism and methodology: Social sciences issues.* Bloomington: Indiana University Press.

Hrdy, S. B. 1999. *Mother nature: Maternal instincts and how they shape human species.* New York: Random House/Ballantine Books.

Kaiser, A., E. Kuenzli, D. Zappatore, and C. Nitsch. 2007. On females' lateral and males' bilateral activation during language production: A fMRI study. *International Journal of Psychophysiology* 63:192–198.

LeVay, S. 1991. A difference in hypothalamic structure between heterosexual and homosexual men. *Science* 253:1034–1037.

McFadden, D. 1993. A masculinizing effect on the auditory systems of human females having male co-twins. *Proceedings of the National Academy of Sciences, U.S.A.* 90: 11900–11904.

———. 2008. What do sex, twins, spotted hyenas, ADHD, and sexual orientation have in common? *Perspectives on Psychological Science* 3 (4): 309–323.

McFadden, D., and C. A. Champlin. 2000. Comparison of auditory evoked potentials in heterosexual, homosexual, and bisexual males and females. *Journal of the Association for Research in Otolaryngology* 1:89–99.

McFadden, D., J. C. Loehlin, and E. G. Pasanen. 1996. Additional findings on heritability and prenatal masculinization of cochlear mechanisms: Click-evoked otoacoustic emission. *Hearing Research* 97:102–119.

McFadden, D., and E. G. Pasanen. 1998. Comparison of the auditory systems of heterosexuals and homosexuals: Click-evoked otoacoustic emissions. *Proceedings of the National Academy of Sciences, U.S.A.* 95:2709–2713.

McFadden, D., E. G. Pasanen, and N. L. Callaway. 1998. Changes in otoacoustic emissions in a transsexual male during treatment with estrogen. *Journal of the Acoustical Society of America* 104:1555–1558.

McFadden, D., and E. Shubel. 2003. The relationships between otoacoustic emissions and relative lengths of fingers and toes in humans. *Hormones and Behavior* 43:421–429.

Rich, A. 1980. Compulsory heterosexuality and lesbian existence. *Signs: Journal of Women in Culture and Society* 5 (4): 631–660.

Spanier, B. 1995a. Biological determinism and homosexuality. *NWSA Journal* 7:54–71. Reprinted in *Same-sex cultures and sexualities: An anthropological reader*, ed. Jennifer Robertson. Malden, MA: Blackwell, 2004.

Spanier, B. 1995b. *Im/partial science: Gender ideology in molecular biology*. Bloomington: Indiana University Press.

Wilson, E. O. 1975. *Sociobiology: The new synthesis*. Cambridge, MA: Harvard University Press.

4

Evaluating Threat, Solving Mazes, and Having the Blues

Gender Differences in Brain-Imaging Studies

CLAUDIA WASSMANN

The aim of this chapter is to highlight gender norms that are embodied in and reinforced, produced, or revised by functional magnetic resonance–imaging (fMRI) research on gender differences. Because assumptions about gender are oftentimes written into the design of a study, some fMRI studies produce or reinforce common gender stereotypes with regard to cognitive functions, such as the claims that women can't read maps or that surgeons are mostly male because they are better at mental rotation. Research finds that women are more "sensitive" and responsive to social cues, more prone to sadness and depression, and less angry than men. These differences are often explained by allegedly evolutionary adaptations or advantages. Some studies give seemingly brain-based explanations for "female brains," which are equated with empathizing, and "male brains," which stand in for synthesizing (Baron-Cohen 2004).

However, the interpretation of differences in brain processes raises a lot of questions. For instance, gender differences are frequently framed in terms of differences in lateralization of brain functions. Research today posits right-brain dominance for men and left-brain dominance for women in terms of emotion processing. Right brain, left brain? Today, these cognitive abilities are being attributed to *opposite* hemispheres and sexes than they were a hundred years ago, when the theory of lateralization of brain function was first discussed. The right-brain–left-brain hypothesis was fashionable at the turn of the twentieth century. At that time, right-brain activity was associated with female, imaginative, emotional, creative thinking in contrast to "left brain logic" (Harrington 1987). Pupils were, for instance, trained to write with their left arms in order to exercise this allegedly subconscious part of their brains. Ironically, if we were to apply the old distinction to the new claims on hemisphere dominance, we would end up with "left-brain logical females" and "right-brain emotional males."

Current brain-imaging research uses gender in multiple ways: Simple gender-discrimination tasks ask the participants in an experiment to determine the gender of a face or to react to an emotional message displayed; imaging paradigms measure gender differences in sexual arousal; and group analyses compare both performance and brain activation in male and female participants during specific activities, such as performing language tasks or solving mathematical equations. Furthermore, brain-imaging studies speak of "gender differences" or "sex differences" somewhat indiscriminately. Sex differences are commonly thought of as being biologically determined, whereas gender differences are supposed to be engendered by social or cultural forces.[1] Nonetheless, in the neuroscience literature, there is often a slippage between the two terms.

A hard distinction between sex and gender, however, can also be thought of as problematic because it is predicated on a deterministic and static view of biology. Yet, the interaction of genes and the environment is a dynamic process. As Melissa Hines put it, "No matter whether hormones or other factors, including social factors, caused us to develop in a certain way, the influences are translated into physical brain characteristics, such as neurons, synapses, and neurochemicals" (2004, 214). The underlying principle of gender studies is that gender is a socially constructed entity and not a natural state. Indeed, Judith Butler (2006) has persuasively problematized the sex/gender dichotomy, and there is ample evidence that gender roles do change over time. From this point of view, research that points out biological differences between men and women is often met with reserve. It seems almost as if research results must be flawed and misguided from the outset. However, as Siri Hustvedt nicely pointed out "except for hermaphrodites, we are born either as a man or as a woman, and biological differences do exist between genders and this does not necessarily mean the repression of one through the other" (2010, 202). Giving birth, for instance, remains a physical event in spite of the varying social constructions that differ among cultures and also change over time. Yet the assertion that biological differences in brain architecture and function allegedly translate to gender-related differences in cognitive performance remains a thorny issue.

In the remainder of this chapter then, I present examples of scientific studies that explore gender differences in brain function and that elucidate the ways in which functional brain-imaging research uses gender. When fMRI became widely available at the turn of the twenty-first century, research using this new technology picked up where previous studies on gender differences had left off. Indeed, differences in mathematical abilities were studied during the 1970s and the 1980s, and many studies on gender differences in emotion appeared during the 1980s and the early 1990s (E. Hall 2000; J. Hall 1978). Therefore, it is worthwhile asking, To what extent are gender norms being perpetuated, enforced, or revised in fMRI studies? And how did these very norms shape the design of brain-imaging research? In particular, I give three sets of examples in the following order: (a) studies on emotion processing, including empathy, depression, and pain; (b) studies on cognitive abilities, such as mathematical and linguistic skills; and (c) proposed functional differences in male and female brains. I begin with an introduction to the popular attempt to find or define the "female brain" (Baron-Cohen 2004; Blum 1997; Brizendine 2006; Darlington 2009; James 2009; Rogers 2001).

Female Brain

Babies are cute. Most women seem to agree on this claim. That is why a group of behavioral biologists and psychiatrists in Germany and the United States studied what underlies the appreciation of the so-called baby schema (Glocker et al. 2009). The baby schema was first defined by the ethologist Konrad Lorenz: A round face with big eyes, high and protruding forehead, chubby cheeks, small nose and mouth. Apparently, baby schema features influence how people rate the cuteness of a baby, and the more infants are perceived as cute prompts more attention to the infant. Infants who showed enhanced baby schema features were rated the cutest and were also rated as smarter, more likeable and cheerful, and more healthy and friendly than other infants. Thus, the study tested women's brain reactions to an enhanced baby schema. The baby features generated activation in parts of the brain that belong to the brain's "reward" system, namely, the mesocorticolimbic system (dopaminergic midbrain, nucleus accumbens, amygdala, and ventromedial prefrontal cortex). In particular, baby schema activated the nucleus accumbens, which is a key structure of the brain's reward system that is linked to the anticipation of reward (O'Doherty 2004). Therefore, the researchers claimed that they have probably discovered the physiological mechanism by which baby schema promotes human care-giving and altruistic behavior regardless of kinship. This would represent an evolutionary advantage to the species, their argument went, because it has the function of enhancing survival of offspring in the group. While the authors discussed a possible cultural and educational component of care-giving behavior, they nonetheless highlighted the strong biological, hardwired physiological core of the observed effect. Although only female volunteers participated in this particular study, earlier studies that have found sex differences in reaction to the baby schema have shown that responses to the perception of infantile cues were stronger in women than men. Another study found that even children at four months of age already demonstrate responsiveness to babyish characteristics (McCall and Kennedy 1980).

Social Bonding

The connection between social bonds and the hormone oxytocin has been the focus of much current research. Oxytocin acts as a neurotransmitter in the brain, and it is also the hormone that is released during orgasm as well as in breastfeeding mothers. Studies in animals pioneered by Thomas Insel have shown that oxytocin plays a key role in complex emotional and social behaviors such as attachment, partner bonding, social recognition, and aggression. More recently, fMRI studies have enabled investigations of the effects of oxytocin in the human brain. Researchers have suggested that oxytocin plays a key role in the development of sexually dimorphic features in the human brain, which are involved in social cognition and lead to observable differences between men's and women's behavior.

For example, neuropsychiatrists at the University of Tokyo have linked altruistic cooperativeness in humans to allegedly female brain structures (Yamasue, Abe, Suga, Yamada, Rogers, et al. 2008). According to the "social-brain hypothesis," the evolution of the human brain reflects the increased necessity for information processing in order to navigate a complex social environment. Taking up the social brain hypothesis, psychologists have argued that enlarged brain size reflects social reciprocity and cooperation. Indeed, Yamasue and colleagues claimed that the development of "social brain" regions is affected by sexually dimorphic factors, which leads to both higher gray matter volumes and higher cooperativeness in women (Yamasue, Abe, Suga, Yamada, Rogers, et al. 2008). Arguably, young adult women showed more cooperativeness than males. And within this framework, the higher cooperative behavior is interpreted as biological by linking it to larger gray matter volumes in the "social brain" regions, which include bilateral posterior inferior frontal and left anterior medial prefrontal cortices. In particular, the research found that female brains showed higher regional gray matter volumes especially in the sexually dimorphic regions of the brain.[2] Thus, the study concluded that the correlation of gray matter volumes and cooperativeness was specific to females. Subsequently, the authors argued that oxytocin should be examined as the "candidate that causes the sexually dimorphic aspect of human social reciprocity [and] social brain development" (Yamasue et al. 2009, 129).[3] In another article, the same authors suggested that there exists yet another allegedly female-specific correlation between brain structures and behavior, namely, higher anxiety-related personality traits and smaller regional brain volume in the left anterior prefrontal cortex (Yamasue, Abe, Suga, Yamada, Inoue, et al. 2008).

Hormones and Brain Development

Gonadal hormones play an important role in early human development. They influence the development of external sexual organs, and they are also thought to influence behavior. Because fetal testes already produce testosterone prenatally (from week eight to week twenty-four of gestation), male fetuses are exposed to higher levels of testosterone than are female fetuses. Research on prenatal androgen exposure now suggests that prenatal testosterone levels influence *postnatal* human behavior, specifically in one manifestation with respect to play attitudes. Furthermore, researchers have found that the early hormone environment influences brain development by determining which nerve cells live or die, which regions of the brain nerve cells interconnect with, and which neurotransmitters are used (Hines 2008). Current information about the impact of early hormone exposure stems mainly from females who suffer from a genetic disorder called congenital adrenal hyperplasia. Girls affected by congenital adrenal hyperplasia experienced higher levels of prenatal androgen exposure, which influences the formation of their external genitals and also shifts their play habits. For instance, studies have documented that these girls prefer "boys' toys," an observation that would counteract the commonly held belief in the social sciences that toy preference results from socialization (Weitz 1977). Brain-imaging studies have also found in these girls an activation pattern of amygdala activity like that found in males during viewing of emotional pictures.

To date, however, no consistent evidence has been established of the effects of the prenatal hormonal environment on cognitive abilities such as visuospatial and mathematical abilities or on verbal fluency, for which gender differences are frequently ascribed. Likewise, research results have been inconclusive regarding the influence of postnatal hormone levels on these cognitive abilities (Hines 2004). In spite of inconclusive research results, however, some neuropsychiatrists go as far as speaking of "the female brain" that purportedly has large resources dedicated to communication and emotion. Allegedly, the hippocampus in females is larger than the hippocampus in male brains, and 11 percent more neurons are dedicated to language. Proponents of the female brain theory suggest that because estrogen is secreted in female infants from age six months to twenty-four months, girls prefer playing with girls; and, because testosterone levels are lower, they are more cooperative, less competitive, and less aggressive in adulthood (e.g., Brizendine 2006). Additionally, it is argued that adult women are later able to use both sides of the brain because their brain circuits for communication develop "unperturbed by" testosterone during fetal development (Brizendine 2006, xviii). Other authors, however, have cast doubt on both the very relevance of these research results and their interpretation. In a review of Brizendine's *The Female Brain*, Rebecca Young and Evan Balaban (2006) pointed out that the differences in brain structure between male and female brains are small and that there is a great deal of overlap. Furthermore, they argued that regulatory feedback loops in the brain make up for any differences. In other words, gender differences are not hardwired.

Obviously, one distinctly female feature in women is the menstrual cycle, so it should come as no surprise that research has also investigated changes in emotional processing across the menstrual cycle. Hormonal changes during the female menstrual cycle are known to transiently affect certain aspects of brain functioning (Andersen and Teicher 2000; Lindamer et al. 1997; Protopopescu et al. 2005). For example, a group of psychiatrists and neuroscientists have studied how the menstrual cycle affects the emotional processing of linguistic stimuli (Protopopescu et al. 2005). That is, they monitored brain reaction to words with positive, negative, or neutral emotional valence in twelve healthy women who had never suffered from premenstrual mood shifts or premenstrual symptoms. Premenstrual mood symptoms are commonly thought to include irritability, tension, depression, loss of control, sleep disturbance, fatigue, food cravings, physical symptoms, and social withdrawal. These symptoms are said to occur most often one to five days before the onset of menses (during the late luteal phase) and they are least severe during days eight through twelve after the onset of menses (the late follicular days). As a result of this view of women's hormonal cycles, the participants in this study underwent two functional magnetic resonance–imaging (fMRI) scans—one brain scan in each phase of the cycle—in order to assess potential differences in brain activity related to hormonal changes. The scans that took place during the luteal phase revealed increased activation of the medial orbitofrontal cortex in reaction to negative words and decreased activity in the lateral orbitofrontal cortex. The pattern was opposite for the follicular days. During the luteal phase decrease of activation also appeared in the left insula and middle cingulate. In considering the neuroanatomic connections of these brain regions, the authors argued that the limbic system might be more excitable premenstrually (luteal phase). That means that when exposed to negative stimuli during this phase of the menstrual cycle, a greater top-down modulation of limbic activity is required, making it harder to shut off negative affect. In other words, due to regularly occurring hormonal changes over the course of the menstrual cycle, women are likely to react more strongly to emotion-eliciting events, especially with regard to negative emotions, and remain for longer periods of time in the grip of emotion.

Evolutionary Explanations

Is there then a "female brain"? How would we know? Gender differences in brain response when viewing erotic pictures have been reported. One fMRI study probed differences in the activation of specific brain circuits in reaction to viewing erotic pictures in male, female, and male-to-female transsexual participants (Gizewski et al. 2009). In this study the authors found no specific activation pattern in male-to-female transsexuals in response to erotic pictures, which means they saw the "female pattern" of brain activation in response to this kind of stimulus material. In another study of affective picture perception that tested gender-specific reactions to erotic pictures, gender differences in the reaction of the visual cortex were interpreted in evolutionary terms (Sabatinelli et al. 2004). The difference was this: while both men and women showed greater activation in the extrastriate visual cortices in response to pictures with negative or positive emotional valence compared to neutral images, men showed greater activation specifically during erotic picture perception. This was interpreted as being due to the motivational relevance of the stimulus material and represented a gender-specific visual mechanism for sexual selection. The authors claimed, "It may be an evolved species survival mechanism, optimizing mate selection (for good health, and youth)" (Sabatinelli et al. 2004, 1111). Or to put it differently, men's brains have evolved to respond to particular features in women that supposedly connote a good reproductive partner. Other studies have demonstrated gender differences in the processing of disgust- and fear-inducing pictures (Schienle 2005). Here, women reacted more strongly to those aversive pictures than men. However, men showed greater activation in bilateral amygdala and left fusiform gyrus, a response pattern that was again interpreted in evolutionary terms: allegedly it reflected "greater attention from males to cues of aggression in their environment" (Schienle 2005, 277).

However, grounding gender-based differences in brain function is not so straightforward. For most aspects of cognitive functions, research results have been equivocal. Sometimes differences in brain function appear only when brain-imaging data are analyzed separately by groups. And sometimes differences that are expected to appear fail to show. For instance, in a study that assumed that men and women "might display at least some distinct characteristics of neurocognitive organization," no sex differences in brain activation showed during object naming of tools versus plants (Garn et al. 2009, 610). The studies mentioned so far elucidate the ways in which gender differences are being investigated using fMRI.

Let me now turn to specific studies that probe common gender-based stereotypes. Previous research had posited that women are more emotional than men; women get sad while men get angry, and girls are afraid whereas boys are disgusted. Gender differences with regard to affective responses were one of the first topics studied with the new brain-imaging techniques (Wager et al. 2003; McClure 2000). Among the most commonly discussed findings are differences concerning the evaluation of threat (McClure et al. 2004), the level of arousal when looking at affective pictures and the appraisal of arousing stimuli (Killgore and Yurgelun-Todd 2001; Klein et al. 2003; Lee et al. 2002), and differences with regard to empathy and depression. In studies, women frequently score higher on standard tests of empathy, social sensitivity, and emotion recognition than do male participants.

Emotion Processing: Empathy, Depression, Pain

Do "female brains" process emotion differently than "male brains"? And is this the reason why women are more vulnerable to depression and to chronic pain conditions? Numerous studies have investigated gender differences with regard to the processing of emotional information, depression, and empathy. These studies often use pictures of facial affect as visual stimulus material in order to elicit emotions. For instance, in a study entitled "Are Emotions Contagious?" male and female participants looked at pictures of happy, sad, or neutral faces taken from Paul Ekman's (Ekman and Friesen 1976) collections of pictures of facial affect. A group of psychiatrists monitored quality, quantity, time course, and gender differences in evoked emotions while viewing these emotionally expressive faces (Wild et al. 2001). Female participants in this study reacted more strongly to these pictures than did men. In their interpretation of the results, the researchers reaffirmed the assumption that women are more emotional than men. The "induction of emotional processes within a subject by the perception of emotionally expressive faces," they argued, "is a powerful instrument in the detection of emotional state in others and is the basis for one's own reactions" (Wild et al. 2001, 109). Along similar lines, a German group of neuropsychologists and neuroscientists investigated whether "gender specific neural mechanisms of emotional social cognition" exist that could account for the fact that women score higher on empathy, social sensitivity, and emotion recognition than men (Schulte-Rüther et al. 2008, 393). Their data suggested that this is indeed the case. Women used "areas containing mirror neurons to a higher degree than males during both SELF- and OTHER-related processing in empathic face-to-face interactions" (Schulte-Rüther et al. 2008, 393). The "female advantage in decoding of non-verbal emotional cues," they argued, translated to differences in regional brain activation in brain networks supporting empathy (393). Males and females seem to rely on different strategies when assessing their own emotions. A study entitled "Are Women Better Mind Readers?" also affirmed a slight female advantage in tasks that required reflection about the mental states of oneself and others (Krach et al. 2009).

Gender differences in emotion processing are frequently framed in terms of differential activation of the two hemispheres of the brain. Indeed, studies suggested opposite laterality effects in amygdala activation in men and women (Hofer et al. 2006). For instance, when male and female participants were asked to put themselves in a positive or a negative mood with the help of black-and-white photographs, right amygdala activation showed in men and left amygdala activation showed in women during negative emotions. The observed effects occurred in different brain regions during positive mood induction and during negative mood induction. In men a laterality effect, meaning activation of only one side, was most marked for negative emotions while the amygdala was activated bilaterally, or both sides, for positive emotions in men. In contrast to men, women seemed to use an integrated network of brain areas in emotion processing.

Another study partially confirmed the observed sex differences for amygdala and hippocampus activation in emotional memory and highlighted the importance of anticipation of aversive events, which might be stronger in women (Mackiewicz et al. 2006). Researchers argue that these differences in emotion processing may undergird gender differences in how men and women deal with emotions. It might also help explain women's greater vulnerability to depression. Gender differences in brain activation in response to viewing fearful and neutral expressions manifest early, with studies showing gender differences already pronounced in children (Thomas et al. 2001). Specifically, researchers observed that boys show a decrease in amygdala activity in response to repeatedly presented fearful faces, whereas girls showed no such attenuation of the amygdala response. The authors hypothesized that this pattern was a possible mechanism that could account for the higher vulnerability of girls to depression during puberty.

Similarly, studies on sadness were among the first conducted with the new brain-imaging techniques. Some of these studies used mood induction paradigms that demonstrate differences in amygdala activation patterns in response to self-induced sadness or happiness. For instance, women "activated a significantly wider portion of their limbic system than did men during transient sadness" even though women reported the same degree of sadness as men when asked how they felt (George et al. 1996, 859). Another early study found activation of the right amygdala in reaction to self-induced sadness only in male participants, not in females (Schneider et al. 2000). The authors interpreted their results as being indicative of "a more focal" processing of sadness in men (Schneider et al. 2000, 226). However, *intra-individual* variability also exists and manifests in brain activation. For instance, in one study in which healthy female volunteers underwent two brain scans at different points in time while experiencing transient sadness induced by viewing sad film excerpts, the authors saw brain activation in the anterior temporal pole and insula during the first scan, while in the second scan sadness was correlated with significant activation in the orbitofrontal and medial prefrontal cortices (Eugène et al. 2003). This indicates that complex feelings such as sadness can be created through different forms of cognitive activity, which are represented by differential brain activation. But it also shows that mapping complex behaviors onto a simple brain activation pattern seen in fMRI is very difficult indeed.

Sadness, Stress, and Depression

It is important to note that gender differences have been found in neural responses to psychological stress. Women, the argument goes, adopt a "tend-and-befriend" response to stress, whereas men are more likely to adopt a "fight-or-flight" response. The tend-and-befriend response means nurturing offspring and affiliating with social groups in order to avoid danger. The fight-or-flight response is characterized by increased focus, alertness, and fear. In a well-designed study that investigated the neurobiological underpinnings of behavioral responses related to psychological stress, sixteen men and women underwent fMRI scans while solving mathematical equations under pressure (Wang et al. 2007). The scans showed marked differences in increase or decrease of brain activity in different brain regions in men and in women in response to subjectively perceived stress. In men, activation increased in the right prefrontal cortex, associated with negative emotions and vigilance, and activation decreased in the left orbitofrontal cortex, associated with positive emotion and hedonistic goals. Additionally, cortisol levels surged, indicating the activity of the physiological stress system. In contrast, the female stress response lasted longer; and in terms of the physiological cortisol response, women seemed to experience more stress than men by low-stress tasks. In the female stress response, activation showed in the limbic system, in brain regions that are part of the neurobiological reward system (ventral striatum, putamen, insula, and cingulate cortex) and that are also associated with perceived anxiety (striatum) and involved in attentional processing of emotion, self-assessment of the mental state, empathy, and reactions to social exclusion (anterior cingulate cortex, ACC). In particular, the lasting stress response in women was characterized by persistent activation of the ACC, which is connected with the amygdala. The authors argued that these data fit well with a cognitive style of ruminative thinking, which seems to be more prevalent in women. The persistent activation of the limbic system and the "lack of containment" of cortisol levels due to "sluggish" feedback have been described as a potential mechanism in the development of depression in women (Wang et al. 2007, 238). In females, cortisol levels were equally high for low and high stress tasks. While there was no absolute difference in cortisol levels between men and women, there was a difference in the delay of shutting down of the cortisol response in women compared to men. Accordingly, some authors have argued that compromised cortisol feedback effects on arousal of the physiological stress axis in women represent a major neurobiological pathway mediating the tendency of women to develop depression (Young and Altemus 2004).

Furthermore, women reacted more strongly in the study than did men to social exclusion as a psychological stressor. This would fit well with the prevalence of adopting a strategy of appeasement instead of a fight-or-flight response in stressful situations. Thus, the study demonstrated marked qualitative and quantitative differences among male and female participants in psychological measures of the stress response, in neuro-imaging data, and in the time pattern of the physiological stress response. Nevertheless, the authors of the study cautioned against interpreting their results along the lines of common stereotypes such as "emotional women" and "rational men." Gender difference in emotionality per se, they argued, "may be an ill-posed question" (Wang et al. 2007, 236). Instead, the differences seen may represent the biological underpinnings of different coping strategies for stress, possibly associated with perceived coping potential or capacity, which can be adopted by both sexes.

Pain

Women suffer more frequently from both depression and chronic pain conditions, especially after menopause. This might be in part related to a higher sensitivity found in women compared to men for information originating in the depth of the body, from the joints, muscles, and viscera (Henderson et al. 2008). These signals from the interior of the body might be more acutely perceived in females or receive more attention. Researchers have claimed this may result in greater pain sensitivity. In terms of brain activation, differences seen between genders (in activation of mid-cingulate cortex, dorsolateral prefrontal cortex, hippocampus, and cerebellar cortex) "may reflect differences in emotional processing of noxious information" (Henderson et al. 2008, 1867).

These findings may help explain why women are more frequently affected by chronic pain conditions. However, research on gender differences in pain sensitivity has also produced mixed results. Both increased and decreased brain activation in several areas of women's brains has been reported, and potential "biopsychosocial mechanisms" have been proposed as an explanation (Fillingim et al. 2009). Over all, though, the balance seems to tip in the direction of higher pain sensitivity in women. Compared to men, women showed "increased self-related attention during anticipation of pain and in response to intense pain" (Straube et al. 2009, 689). Indeed, female study participants experienced mild and moderate pain at lower stimulus intensity than males. Furthermore, pain stimulation from very low to very high intensity generated stronger activation in a region of the medial prefrontal cortex in women. This region of the brain has been implicated in introspective and self-focused information processing.

Taken together, then, these studies demonstrated some structural and functional differences in emotion processing, which were interpreted along gender lines and used to explain, in part, the observed higher rates of depression in women. Various other explanations have been given to account for the observation that depression is more frequent in females (Hamann 2005; Hasler et al. 2004). For example, Laura Hirshbein (2006) even argued that gender played an important role in the very creation of "depression" as a diagnostic category.

Cognitive Abilities: Mathematics, Mental Rotation, Language

Purported differences in cognitive strategies between men and women are highly contentious. Such differences in brain functioning fascinate both researchers and the lay press (James 2009). Differences have been reported with regard to spatial navigation, mathematical abilities, and language processing. What exactly do these biological differences mean? One of the most common stereotypes is that women are bad in math and therefore underrepresented in scientific professions (Rosser 2008). The theory on gender differences in mathematical skills advantaging males dates back a hundred years (Shields 1975). Functional brain-imaging studies on mathematical abilities, however, have produced ambiguous results, and the interpretation of earlier data has changed as well. While newer studies tend to see equality, older studies confirmed the alleged superiority of males in mathematical reasoning. Indeed, an older positron emission tomography (PET) study found differences in brain activation during mathematical reasoning on a task similar to the Scholastic Aptitude Test (SAT), with rising activity in the right temporal lobes in men only, which would represent the neurophysiological equivalent of better task performance (Larson et al. 1995). In contrast, a recent study interpreted gender differences in the opposite direction. Differences in the functional and structural neuro-anatomy of mathematical cognition could represent a "more efficient use of neural processing resources in females" (Keller and Menon 2009, 342).

Similarly, a 1990 meta-analysis of data from studies conducted during the 1970s and 1980s on mathematical abilities revealed major differences in math performance between men and women when more complex mathematical problems had to be solved. In contrast, more recent studies have argued that gender similarities rather than differences characterized math performance (Hyde et al. 2008). The authors explained their findings as the result of a shift in learning behavior. Gender differences have disappeared because girls are now taking calculus classes in high school. Moreover, a recent cross-national study has confirmed that while more males than females were found in the group of highest mathematical abilities, some women did belong to the group of highest abilities, and differences in mathematical performance significantly correlated with sociocultural factors, such as gender gaps in the attribution of resources, parental income, the accessibility of education, and the quality of mathematical education (Hyde and Mertz 2009).

Spatial Navigation and Mental Rotation

Next to math skills, spatial navigation counts as a prominent candidate for gender differences turned into gender stereotypes. Why women "can't read a map" or why surgeons are mostly male gets explained as differences in brain structure and function. But what exactly does it mean if scientists state: "Gender-specific group analysis revealed distinct activation of the left hippocampus in males, whereas females consistently recruited right parietal and right prefrontal cortex. Thus we demonstrate a neural substrate of well established human gender differences in spatial-cognition performance" (Grön et al. 2000, 404)? Indeed, previous behavioral studies have shown that during spatial navigation women rely on landmark cues whereas men use both geometric cues and landmark cues (Linn and Petersen 1985). Accordingly, the differences in brain activation found in brain-imaging studies were interpreted in terms of different cognitive strategies employed by men and women in order to get to the same results, which were, in the case of women, viewed by researchers as "more costly." The pattern of activation seen more frequently in women (with activation of the prefrontal cortex), arguably, reflects the demand on working memory to hold information on landmark cues "online," which requires more nervous-processing resources. In contrast, men deployed more economical strategies. First, higher activation showed in the hippocampus, enabling parallel processing of multiple geometrical cues. Furthermore, the activity of the hippocampus suggested the reliance on episodic memory rather than on working memory in spatial navigation. Thus, the activation of the left hippocampus in men and of the right frontoparietal cortex in women during spatial navigation would reflect a "gender specific" recruitment of nervous resources that "differentiate male from female subjects" (Grön et al. 2000, 407). In this model, gender-different neural networks would be the substrate of differences in performance, meaning that allegedly an inferior performance in spatial cognition in women is predicated on the very wiring of their brains.

So is there, then, a fundamental gender difference in cognitive performance? This does not seem to be a certainty. Gender-specific cortical activation patterns during mental rotation tasks, which test a kind of spatial navigation, have confirmed different activation patterns in female and male brains despite similar performance (Jordan et al. 2002). Same results, but different pathways to get there? Sex differences in cognitive abilities such as general intelligence have been shown to be negligible. The greatest differences have, indeed, been found with regard to visuospatial abilities such as mental rotation tasks (Thomas and Kail 1991; Butler et al. 2006). But here again test results have varied from small to large differences. Results are often task specific; for instance, the largest difference between men and women, favoring adult males, appeared in mental rotation tasks that required rapid and precise mental rotation, whereas other spatial visualization tasks showed negligible sex differences (Voyer et al. 1995).

Gender differences have manifested in men's and women's "spatial perspective-taking" ability, which is defined as the ability to translocate one's own egocentric viewpoint to somebody else's viewpoint in space. In one study, males were more likely to employ an object-based strategy in contrast to an "egocentric perspective transformation," which was consistently employed in females (Kaiser et al. 2008). In other words, men and women used different frames of reference. When women had to describe a scene from a third-person perspective, they virtually put themselves in the place of the other agent, adopting the gaze of the other, whereas men represented spatial relations of objects independently of their own location. Therefore, women had to do mental rotation, whereas men did not. The authors found that women's performance had a pronounced decline as the complexity of the perspective-taking task increased; in contrast, men's performance scores did not decline as dramatically. The authors argued that the differential recruitment of brain regions seen in their fMRI data most likely reflected different strategies in solving the spatial perspective-taking task.

Language

Although women are bad in math, they excel in verbal abilities, so the stereotypes have it. However, even fMRI analyses of linguistic skills have produced mixed results. Large studies have been conducted that demonstrate sex differences in verbal abilities within the normal population. Yet a critical review of putative sex differences in verbal abilities and language cortex could not confirm substantial differences. A careful reading of the research results suggests that gender differences in language proficiency do not exist. A slight early advantage for girls in language acquisition seems to gradually disappear with age. A difference in language lateralization in adults, which has been suggested, was not substantiated by evidence. Language processing seems to be strongly left lateralized in both sexes and "substantive differences" do not seem to exist between men and women in the "large-scale organization of language processes" (Frost et al. 1999, 199). In addition, overall results from studies on regional gray matter distribution (using voxel-based morphometry) have indicated "no consistent differences between males and females in language-related cortical regions" (Wallentin 2009, 175).

Nevertheless, the desire to find gender differences seems to be great. Some authors have pointed out "dramatic sex differences in the pattern of brain-behavior correlations" that, according to their reading, reflected "fundamental differences in the nature of processing required for accurate performance" in language processing (Burman et al. 2008, 1359). Indeed, in their study on sex differences in neural processing of language among children aged nine to fifteen years, the authors found that "girls rely on a supramodal language network" and boys process visual and auditory words differently (Burman et al. 2008, 1349). These differences, however, may be lost in adulthood. A study on silent lipreading in normal-hearing male and female subjects suggested that "females associated the visual speech image with the corresponding auditory speech sound, whereas males focused more on the visual image itself" (Ruytjens et al. 2006, 1835). Along the same lines, a French study argued, based on brain activity pattern, that men use mental-imaging strategies in word retrieval (Gauthier et al. 2009). Nonetheless, the authors conceded that these strategies can be used by both sexes; they seem to be "not specific to men but more usual in men than in women" (Gauthier et al. 2009, 168).

Basic Brain Functions

Sex differences in the human brain have been pointed out for most sensory systems: the olfactory system, the auditory system, and the visual cortex. The human olfactory system is depicted as a sexually dimorphic network. Here, women have shown, on average, greater activation of the left and right inferior frontal regions and scored better than men on psychological odor identification tests (Garcia-Falgueras et al. 2006; Yousem et al. 1999). In the human primary visual cortex, men exhibited a greater increase of activation in reaction to blue light. This was interpreted as a sex-related difference in central nervous dopamine function (Cowan et al. 2000). Studies on frequency-dependent activation of the visual cortex have suggested that observed gender differences in working memory function may be associated with a more global information processing, with right hemisphere dominance particularly in men (Kaufmann et al. 2001).

As I mentioned earlier, widely discussed gender-based differences concern the lateralization of brain functions, in particular, with regard to amygdala activation, especially in negative emotions, where right hemisphere dominance was suggested for men, and left brain dominance for women (Kempton et al. 2009). Gender-related differences in lateralization were also found with regard to the functional organization of the brain for working memory (Speck 2000). "Highly significant differences" exist in working memory organization, the authors claimed (Speck 2000, 2581). While men showed bilateral or right-sided activation (of the prefrontal cortex, parietal cortex, and caudate), women showed predominantly left hemisphere activation, higher task performance accuracy, and slightly slower reaction times. Sex differences showed also in sensory gating of the thalamus during auditory interference of a visual attention task (Tomasi et al. 2008). Women had "less suppression or gating of auditory evoked response than men" (Hetrick et al. 1996). This finding suggests less neuronal response, researchers claimed, and could explain why women have a larger startle response (Kofler et al. 2001) and perceive unexpected noise as louder than men do (Kimura 1999). Laterality effects in the opposite direction were found in the auditory system. While women showed a strong left ear dominance with information processing in the right hemisphere, men showed a right ear dominance with information processing in the left hemisphere (Lewald 2004).

Conclusion

What can we make of all these examples of biological differences between the sexes that contemporary neuroscience has generated? As evidence of gender differences accumulates in brain-imaging research, it is important to note that, on the one hand, real differences might not be detectable with today's techniques and, on the other hand, remarkable similarities between the sexes far outnumber the differences (Wager and Ochsner 2005). Indeed, studies have claimed both that sex differences in the brain translate into cognitive differences and that women use their brains differently to get the same results (Goldstein et al. 2005). Several recent reviews of fMRI studies on gender-related differences call for caution in the interpretation of brain-imaging data. Gender differences should be seen much more critically because of "paradigmatic, methodological and statistical defaults" that interfere with assessing the presence or absence of sex/gender differences (Kaiser et al. 2009, 49; Kaiser et al. 2007). Moreover, dealing with the sex/gender variable will "inevitably lead to the detection of differences rather than to the detection of similarities" (Kaiser et al. 2009, 49). What is more, some studies have begun questioning the impact of differences that brain-imaging studies reveal altogether. Critics have pointed out that "virtually all differences in brain structure, and most differences in behavior, are characterized by small average differences" (Young and Balaban 2006, 634). While sex differences in the human brain do exist, there is no straightforward relationship to gender differences in human behavior. Few data are available that link sex differences in brain structure to functional differences (Hines 2004, 211). As a review of the popular book *The Female Brain* (Brizendine 2006) suggested, we should stay clear of doing "psychoneuroindoctrinology" (Young and Balaban 2006).

NOTES

1. For a discussion of the sex/gender distinction from the social sciences' point of view, see Nye 2008.
2. For a critical introduction into research on sexual dimorphisms, see Einstein 2007.
3. The researchers further claimed that oxytocin should also be examined in regard to the pathogenesis of autism spectrum disorder because autism is rare in females (Yamasue et al. 2009).

REFERENCES

Andersen, S. L. and M. H. Teicher. 2000. Sex differences in dopamine receptors and their relevance to ADHD. *Neuroscience and Biobehavioral Reviews* 24:137–141.

Baron-Cohen, S. 2004. *The essential difference: Male and female brains and the truth about autism.* New York: Basic Books.

Blum, D. 1997. *Sex on the brain: The biological differences between men and women.* New York: Viking.

Brizendine, L. 2006. *The female brain.* New York: Morgan Road Books.

Burman, D. D., T. Bitanc, and J. R. Booth. 2008. Sex differences in neural processing of language among children. *Neuropsychologia* 46 (5): 1349–1362.

Butler, J. 2006. *Gender trouble: Feminism and the subversion of identity.* New York: Routledge.

Butler, T., J. Imperato-McGinleyb, H. Pana, D. Voyerc, J. Corderob, Y.–S. Zhub, E. Sterna, and D. Silbersweig. 2006. Sex differences in mental rotation: Top-down versus bottom-up processing. *NeuroImage* 32:445–456.

Cowan, R. L., B. Frederick, M. Rainey, J. M Levin, L. C. Maas, J. Bang, J. Hennen, S. E. Lukas, and P. F. Renshaw. 2000. Sex differences in response to red and blue light in human primary visual cortex: A bold fMRI study. *Psychiatry Research–Neuroimaging* 100 (3): 129–138.

Darlington, C. L. 2009. *The female brain.* Boca Raton, LA: CRC Press.

Einstein, G., ed. 2007. *Sex and the brain.* Cambridge, MA: MIT Press.

Ekman, P., and W. V. Friesen. 1976. *Pictures of facial affect.* Palo Alto, CA: Consulting Psychologists Press.

Eugène, F., J. Lévesque, B. Mensour, J.-M. Leroux, G. Beaudoin, P. Bourgouin, and M. Beauregard. 2003. The impact of individual differences on the neural circuitry underlying sadness. *NeuroImage* 19 (2): 354–364.

Fillingim, R. B., C. D. King, M. C. Ribeiro-Dasilva, B. Rahim-Williams, and J. L. Riley III. 2009. Sex, gender, and pain: A review of recent clinical and experimental findings. *Journal of Pain* 10 (5): 447–485.

Frost, J. A., J. R. Binder, J. A. Springer, and T. A. Hammeke. 1999. Language processing is strongly left lateralized in both sexes: Evidence from functional MRI. *Brain* 122 (2): 199–208.

Garcia-Falgueras, A., C. Junque, M. Giménez, X. Caldú, S. Segovia, and A. Guillamon. 2006. Sex differences in the human olfactory system. *Brain Research* 1116:103–111.

Garn, C. L., M. D. Allen, and J. D. Larsen. 2009. An fMRI study of sex differences in brain activation during object naming. *Cortex* 45 (5): 610–618.

Gauthier, C. T., M. Duyme, M. Zanca, and C. Capron. 2009. Sex and performance level effects on brain activation during a verbal fluency task: A functional magnetic resonance imaging study. *Cortex* 45 (2): 164–176.

George, M. S., T. A. Kettera, P. I. Parekha, P. Herscovitchb, and R. M. Posta. 1996. Gender differences in regional cerebral blood flow during transient self-induced sadness or happiness. *Biological Psychiatry* 40 (9): 859–871.

Gizewski, E. R., E. Krause, M. Schlamann, F. Happich, M. E. Ladd, M. Forsting, and W. Senf. 2009. Specific cerebral activation due to visual erotic stimuli in male-to-female transsexuals compared with male and female controls: An fMRI study. *Journal of Sexual Medicine* 6 (2): 440–448.

Glocker, M. L., D. D. Langleben, K. Ruparel, J. W. Loughead, J. N. Valdez, M. D. Griffin, N. Sachser, and R. C. Gur. 2009. Baby schema modulates the brain reward system in nulliparous women. *Proceedings of the National Academy of Sciences of the United States of America* 106 (22): 9115–9119.

Goldstein, J. M., M. Jerram, R. Poldrack, R. Anagnoson, H. C. Breiter, N. Makris, J. M. Goodman, M. T. Tsuang, and L. J. Seidman. 2005. Sex differences in prefrontal cortical brain activity during fMRI of auditory verbal working memory. *Neuropsychology* 19 (4): 509–519.

Grön, G., A. P. Wunderlich, M. Spitzer, R. Tomczak, and M. W. Riepe. 2000. Brain activation during human navigation: Gender-different neural networks as substrate of performance. *Nature Neuroscience* 3 (4): 404–408.

Hall, E. J. 2000. Developing the gender relations perspective: The emergence of a new conceptualization of gender in the 1990s. *Current Perspectives in Social Theory* 20:91–123.

Hall, J. A. 1978. Gender effects in decoding nonverbal cues. *Psychological Bulletin* 85 (4): 845–857.

Hamann, S. 2005. Blue genes: Wiring the brain for depression. *Nature Neuroscience* 8 (6): 701–703.

Harrington, A. 1987. *Medicine, mind, and the double brain: A study in nineteenth-century thought.* Princeton, NJ: Princeton University Press.

Hasler, G., W. C. Drevets, H. K. Manji, and D. S. Charney. 2004. Discovering endophenotypes for major depression. *Neuropsychopharmacology* 29 (10): 1765–1781.

Henderson, L. A., S. C. Gandeviab, and V. G. Macefield. 2008. Gender differences in brain activity evoked by muscle and cutaneous pain: A retrospective study of single-trial fMRI data. *NeuroImage* 39 (4): 1867–1876.

Hetrick, W. P., C. A. Sandman, W. E. Bunney Jr., Y. Jin, S. G. Potkin, and M. H. White. 1996. Gender differences in gating of the auditory evoked potential in normal subjects. *Biological Psychiatry* 39:51–58.

Hines, M. 2004. *Brain gender.* New York: Oxford University Press.

———. 2008. Early androgen influences on human neural and behavioural development. *Early Human Development* 84 (12): 805–807.

Hirshbein, L. D. 2006. Science, gender, and the emergence of depression in American psychiatry, 1952–1980. *Journal of the History of Medicine and Allied Sciences* 61 (2): 187–216.

Hofer, A., C. M. Siedentopf, A. Ischebeck, M. A. Rettenbacher, M. Verius, S. Felber, and W. W. Fleischhacker. 2006. Gender differences in regional cerebral activity during the perception of emotion: A functional MRI study. *NeuroImage* 32 (2): 854–862.

Houdé, O., L. Zago, F. Crivello, S. Moutier, A. Pineau, B. Mazoyer, and N. Tzourio-Mazoyer. 2001. Access to deductive logic depends on a right ventromedial prefrontal area devoted to emotion and feeling: Evidence from a training paradigm. *NeuroImage* 14 (6): 1486–1492.

Hustvedt, S. 2010. *The shaking woman, or, a history of my nerves.* New York: Henry Holt.

Hyde, J. S., S. M. Lindberg, M. C. Linn, A. B. Ellis, and C. C. Williams. 2008. Diversity: Gender similarities characterize math performance. *Science* 321:494–495.

Hyde, J. S., and J. E. Mertz. 2009. Gender, culture, and mathematics performance. *Proceedings of the National Academy of Sciences of the United States of America* 106 (22): 8801–8807.

James, A. N. 2009. *Teaching the female brain: How girls learn math and science.* Thousand Oaks, CA: Corwin Press.

Jordan, K., T. Wüstenberga, H.-J. Heinzeb, M. Petersc, and L. Jäncke. 2002. Women and men exhibit different cortical activation patterns during mental rotation tasks. *Neuropsychologia* 40 (13): 2397–2408.

Kaiser, A., S. Hallerb, S. Schmitzd, and C. Nitsch. 2009. On sex/gender related similarities and differences in fMRI language research. *Brain Research Reviews* 61:49–59.

Kaiser, A., E. Kuenzli, D. Zappatore, and C. Nitsch. 2007. On females' lateral and males' bilateral activation during language production: A fMRI study. *International Journal of Psychophysiology* 63 (2): 192–198.

Kaiser, S., S. Walther, E. Nennig, K. Kronmüller, C. Mundt, M. Weisbrod, C. Stippich, and Kai Vogeley. 2008. Gender-specific strategy use and neural correlates in a spatial perspective taking task. *Neuropsychologia* 46 (10): 2524–2531.

Kaufmann, C., G.-K. Elbel, C. Gössl, B. Pütz, and D. P. Auer. 2001. Frequency dependence and gender effects in visual cortical regions involved in temporal frequency dependent pattern processing. *Human Brain Mapping* 14:28–38.

Keller, K., and V. Menon. 2009. Gender differences in the functional and structural neuroanatomy of mathematical cognition. *NeuroImage* 47:342–352.

Kempton, M. J., M. Haldane, J. Jogia, T. Christodoulou, J. Powell, D. Collier, S.C.R. Williams, and S. Frangou. 2009. The effects of gender and COMT Val158Met polymorphism on fearful facial affect recognition: A fMRI study. *International Journal of Neuropsychopharmacology* 12 (3): 371–381.

Killgore, W.D.S., and D. A. Yurgelun-Todd. 2001. Sex differences in amygdala activation during the perception of facial affect. *NeuroReport* 12 (11): 2543–2547.

Kimura, D. 1999. *Sex and cognition.* Cambridge, MA: MIT Press.

Klein, S., M. N. Smolka, J. Wrase, S. M. Gruesser, K. Mann, D. F. Braus, and A. Heinz. 2003. The influence of gender and emotional valence of visual cues on fMRI activation in humans. *Pharmacopsychiatry* 36 (suppl. 3): S191–S194.

Kofler, M., J. Müller, L. Reggiani, and J. Valls-Solé. 2001. Influence of gender on auditory startle responses. *Brain Research* 921:206–210.

Krach, S., I. Blümel, D. Marjoram, T. Lataster, L. Krabbendam, J. Weber, J. van Os, and T. Kircher. 2009. Are women better mindreaders? Sex differences in neural correlates of mentalizing detected with functional MRI. *BMC Neuroscience* 10:Article 9.

Larson, G. E., R. J. Haier, L. LaCasse, and K. Hazen. 1995. Evaluation of a "mental effort" hypothesis for correlations between cortical metabolism and intelligence. *Intelligence* 21 (3): 267–278.

Lee, T.M.C., H.-L. Liu, R. Hoosain, W.-T. Liao, C.-T. Wu, K.S.L. Yuen, C.C.H. Chan, P. T. Fox, and J.-Ho. Gao. 2002. Gender differences in neural correlates of recognition of happy and sad faces in humans assessed by functional magnetic resonance imaging. *Neuroscience Letters* 333 (1): 13–16.

Lewald, J. 2004. Gender-specific hemispheric asymmetry in auditory space perception. *Cognitive Brain Research* 19 (1): 92–99.

Lindamer, L. A., J. B. Lohr, M. J. Harris, and D. V. Jeste. 1997. Gender, estrogen, and schizophrenia. *Psychopharmacology Bulletin* 33 (2): 221–228.

Linn, M. C., and A. C. Petersen. 1985. Emergence and characterization of sex differences in spatial ability: A meta-analysis. *Child development* 56 (6): 1479–1498.

Mackiewicz, K. L., I. Sarinopoulos, K. L. Cleven, and J. B. Nitschke. 2006. The effect of anticipation and the specificity of sex differences for amygdala and hippocampus function in emotional memory. *Proceedings of the National Academy of Sciences of the United States of America* 103 (38): 14200–14205.

McCall, R. B., and C. B. Kennedy. 1980. Attention of four-month infants to discrepancy and babyishness. *Journal of Experimental Child Psychology* 29 (2): 189–201.

McClure, E. B. 2000. A meta-analytic review of sex differences in facial expression processing and their development in infants, children, and adolescents. *Psychological Bulletin* 126 (3): 424–453.

McClure, E. B., C. S. Monk, E. E. Nelson, E. Zarahn, E. Leibenluft, R. M. Bilder, D. S. Charney, M. Ernst, and D. S. Pine. 2004. A developmental examination of gender differences in brain engagement during evaluation of threat. *Biological Psychiatry* 55 (11): 1047–1055.

Nye, R. A. 2008. The biosexual foundations of our modern concept of gender. In *Sexualized brains,* ed. N. C. Karafyllis and G. Ulshöfer, 69–80. Cambridge, MA: MIT Press.

O'Doherty, J. P. 2004. Reward representations and reward-related learning in the human brain: Insights from neuroimaging. *Current Opinion in Neurobiology* 14 (6): 769–776.

Protopopescu, X., H. Pan, M. Altemus, O. Tuescher, M. Polanecsky, B. McEwen, D. Silbersweig, and E. Stern. 2005. Orbitofrontal cortex activity related to emotional processing changes across the menstrual cycle. *Proceedings of the National Academy of Sciences of the United States of America* 102 (44): 16060–16065.

Rogers, L. J. 2001. *Sexing the brain.* New York: Columbia University Press.

Rosser, S. V. 2008. *Women, science, and myth: Gender beliefs from antiquity to the present.* Santa Barbara, CA: ABC-CLIO.

Ruytjens, L., F. Albers, P. Van Dijk, H. Wit, and A. Willemsen. 2006. Neural responses to silent lipreading in normal hearing male and female subjects. *European Journal of Neuroscience* 24 (6): 1835–1844.

Sabatinelli, D., T. Flaisch, M. M. Bradley, J. R. Fitzsimmons, and P. J. Lang. 2004. Affective picture perception: Gender differences in visual cortex? *NeuroReport* 15 (7): 1109–1112.

Schienle, A. 2005. Gender differences in the processing of disgust- and fear-inducing pictures: An fMRI study. *NeuroReport* 16 (3): 277–280.

Schneider, F., U. Habel, C. Kessler, J. B. Salloum, and S. Posse. 2000. Gender differences in regional cerebral activity during sadness. *Human Brain Mapping* 9 (4): 226–238.

Schulte-Rüther, M., H. J. Markowitsch, N. J. Shah, G. R. Fink, and M. Piefke. 2008. Gender differences in brain networks supporting empathy. *NeuroImage* 42 (1): 393–403.

Shields, S. A. 1975. Functionalism, Darwinism, and the psychology of women: A study in social myth. *American Psychologist* 30:739–754.

Speck, O. 2000. Gender differences in the functional organization of the brain for working memory. *NeuroReport* 11 (11): 2581–2585.

Straube, T., S. Schmidt, T. Weiss, H.-J, Mentzel, and W.H.R. Miltner. 2009. Sex differences in brain activation to anticipated and experienced pain in the medial prefrontal cortex. *Human Brain Mapping* 30 (2): 689–698.

Thomas, H., and R. Kail. 1991. Sex differences in speed of mental rotation and the X-linked genetic hypothesis. *Intelligence* 15:17–32.

Thomas, K. M., W. C. Drevets, P. J. Whalen, C. H. Eccard, R. E. Dahl, N. D. Ryan, and B. J. Casey. 2001. Amygdala response to facial expressions in children and adults. *Biological Psychiatry* 49 (4): 309–316.

Tomasi, D., L. Chang, E. C. Caparelli, and T. Ernst. 2008. Sex differences in sensory gating of the thalamus during auditory interference of visual attention tasks. *Neuroscience* 151 (4): 1006–1015.

Voyer, D., S. Voyer, and M. P. Bryden. 1995. Magnitude of sex differences in spatial abilities: A meta-analysis and consideration of critical variables. *Psychological Bulletin* 117 (2): 250–270.

Wager, T. D., and K. N. Ochsner. 2005. Sex differences in the emotional brain. *NeuroReport* 16 (2): 85–87.

Wager, T. D., K. L. Phan, I. Liberzon, and S. F. Taylor. 2003. Valence, gender, and lateralization of functional brain anatomy in emotion: A meta-analysis of findings from neuroimaging. *NeuroImage* 19 (3): 513–531.

Wallentin, M. 2009. Putative sex differences in verbal abilities and language cortex: A critical review. *Brain and Language* 108 (3): 175–183.

Wang, J., M. Korczykowski, H. Rao, Y. Fan, J. Pluta, R. C. Gur, B. S. McEwen, and J. A. Detre. 2007. Gender difference in neural response to psychological stress. *Social Cognitive and Affective Neuroscience* 2 (3): 227–239.

Weitz, S. 1977. *Sex roles: Biological, psychological, and social foundations.* New York: Oxford University Press.

Wild, B., M. Erb, and M. Bartels. 2001. Are emotions contagious? Evoked emotions while viewing emotionally expressive faces: Quality, quantity, time course, and gender differences. *Psychiatry Research* 102 (2): 109–124.

Yamasue, H., O. Abe, M. Suga, H. Yamada, H. Inoue, M. Tochigi, M. Rogers, S. Aoki, N. Kato, and K. Kasai. 2008. Gender-common and -specific neuroanatomical basis of human anxiety-related personality traits. *Cerebral Cortex* 18:46–52.

Yamasue, H., O. Abe, M. Suga, H. Yamada, M. Rogers, S. Aoki, N. Kato, and K. Kasai. 2008. Sex-linked neuroanatomical basis of human altruistic cooperativeness. *Cerebral cortex* 18:2331–2340.

Yamasue, H., H. Kuwabara, Y. Kawakubo, and K. Kasai. .2009. Oxytocin, sexually dimorphic features of the social brain, and autism. *Psychiatry and Clinical Neurosciences* 63 (2): 129–140.

Young, E. A., and M. Altemus. 2004. Puberty, ovarian steroids, and stress. *Annals of the New York Academy of Sciences* 1021: 124–133.

Young, R. M., and E. Balaban. 2006. Psychoneuroindoctrinology. *Nature* 443: 634.

Yousem, D. M., J. A. Maldjian, F. Siddiqi, T. Hummel, D. C. Alsop, R. J. Geckle, W. B. Bilker, and R. L. Doty. 1999. Gender effects on odor-stimulated functional magnetic resonance imaging. *Brain Research* 818 (2): 480–487.

PART TWO

Animal Obsessions

5

Telling the Rat What to Do

Laboratory Animals, Science, and Gender

LYNDA BIRKE

To the chagrin of the male chauvinist, the female rat is the more exploratory, the more adventuresome of the sexes. . . . But to the despair of the ultrafeminist, the typical female mouse or monkey is undeniably less aggressive than her male counterpart.

–Mariette Nowak, *Eve's Rib: A Revolutionary New View of the Female*

Rats . . . depend on instinct for a great deal of their behavior, including sexual behavior. They don't learn about the birds and the bees from their folks. . . . So somewhere in the rat brain must be some sort of sex circuits that tell the rat what to do.

–Robert Pool, *Eve's Rib: Searching for the Biological Roots of Sex Differences*

Other animals provide us with myriad metaphors and mirrors of our own behavior. We may at times see nobility in the characters of other animals—or parallels with our own mindlessness. Sometimes, animals seem to provide mirrors of our own society, seeming to behave in ways that fit with particular expectations of gender, for example, as Mariette Nowak (1980) implies. Often, too, the behavior of other animals is used to indicate that something is fixed, instinctive, hardwired into the brain—circuits "telling the rat what to do."

Much of what we know about animal behavior, however, comes from scientific investigations—observations of wild animals—or studies carried out in laboratories, including those assessing rat sexual behavior. In this chapter, I will explore how scientific claims about gender differences relate to how we think about nonhuman animals. Can drawing conclusions from laboratory animals tell us much about human gender or sexuality, or do they tell us only about those animals? What could studies of other animals tell us about complex cultural—human—processes? How could they be pertinent to feminist debates about how gender is constructed? Human observers of free-living wild animals do not simply watch; they bring to their observations their own cultural history and values (see Haraway 1989 with regard to primatology). Among other things, this involves specific expectations of how males and females behave.

Rather than focusing on gender and interpretations of animal behavior in the wild, however, I want to deal here with studies using animals in laboratories. Lab animals, usually primates or rodents, are employed in experimental protocols to test a range of hypotheses about how differences between females and males develop or about the etiology of sexuality.[1] Here, too, cultural expectations of gender or sexual behavior shape how experiments are done and the results that are obtained. So, the second part of the chapter will be, in part, a lab animal ethnography—following animals into laboratory spaces—from which I will argue that how lab animals are kept and how they live their lives also contribute to what might be called a standardization of gender within scientific thought. Finally, in the last section, I will return to the more general question of feminist approaches to thinking about animals, to end with a focus on possibilities rather than limitations.

Reading Gender onto Nature

Our relationship with all kinds of animals is deeply embedded in our unconscious, showing itself in our mythologies, our history, our literatures. In observing the enormous richness and variety in the natural world, we make assumptions and have expectations of what other species are or can do. These assumptions are grounded partly in the history of our culture's broader attitudes toward animals or toward nature in general—we inherit, for example, biblical beliefs in human dominion over the rest of nature, which color our understanding, overlain with a greater sensibility toward nature that arose from the eighteenth century onward. Beliefs about animals are also grounded in our scientific worldview—for not only does science define what different species of animals are and how they came into being, but it also uses them experimentally.

Humans tend to interpret the behavior of other species within frameworks derived from our own experiences, within our specific cultures. Scientists are not exempt from this: it is almost impossible not to bring one's own experiences into observations of animals, however much scientists are trained to pursue an "objective" distance. So, given that we live in a gender-stratified culture, gender divisions are likely to be part of how we see the rest of nature.[2] Indeed, that is the central plank of feminist criticism—that human gender divisions are too easily read onto other animals, while animal societies are too readily held up as mirrors of those divisions.

Looking at animal societies as mirrors of human behavior can be problematic, however. How clearly do the observers really see what distinguishes that species and its behavior? And how much do they interpret the animals in terms of human social organization—including gender divisions? Related to this are widespread cultural beliefs that the behavior of nonhuman animals is instinctive, hardwired, not subject to change. And this is why feminists have particularly resisted extrapolations from animal species to humans: not only does the linkage so often imply mindlessness or (women's) irrationality, but it also leads to claims that gender is built into our biology, which seem to fly in the face of feminist demands for change and our beliefs in the flexibility of human behavior.[3]

In part, feminist concerns about extrapolation from other species resulted from a long history of the use of animal terms to denigrate women; but it also arose because descriptions of gender difference in other species too often lead to claims that gender difference in people is *rooted* in biology. In the early years of second-wave feminism, the primary response to such allegations was to reject them, emphasizing instead the considerable extent to which gender in people was socially constructed. Women's movements were (and are) concerned with issues to do with legal status, social power, equality at work—issues which might be challenged and changed. Rejecting biological claims was, and remains, important as an antidote to claims that gender stereotypes are fixed into the body. A few years ago, for instance, I read about a claim that women were somehow biologically predisposed to like ironing—a curious supposition for those millions of women (like me) who loathe it. More recently, a furor ensued when the president of Harvard University, Lawrence Summers, announced that biological differences explained the relative dearth of women in the physical sciences and engineering. Such biological determinism seldom helps anyone.

A cornerstone of feminist writing for many years has been a distinction between gender and sex (see Zurbriggen and Sherman 2007). "Sex" was taken to mean the bodily distinction between male and female—biological sex—including differences in chromosomes or reproductive anatomy. "Gender," by contrast, was taken to mean sociocultural construction—how we come to be or behave differently, for example. Importantly, the sex-gender distinction served to emphasize the complex ways in which we might acquire our understanding of gender and opened up the possibility that gender was better understood as multiple and fluid. This was a particularly important point for feminist analysis as it allowed understanding of how gender could be produced and reenacted. It also allowed for discrepancy between bodily sex and a person's sense of their own gender, that is, that experiencing oneself as a specific gender did not necessarily map onto being a particular sex.

The sex-gender distinction, however, has also proved problematic for feminism. In particular, it posits biological sex as the *basis* for gender and so inextricably linked to it causally; and it creates a separation between the biological and the social. Because of this focus on the distinction, for a long time feminist theory paid too little heed to the inner workings of the (biological) body and, simultaneously, paid little heed to thinking about animals. It was as though animals had only sex/bodies; people had gender and mind.[4]

The biomedical sciences have often reported sex/gender differences, from muscle strength to biochemistry of cellular function—a search now intensifying with the creation of new fields such as "gender-specific medicine."[5] Seeking differences, however, only serves to reinforce them. When Rhoda Unger first proposed the sex-gender distinction thirty years ago, she pointed out that "men and women are especially alike in their beliefs about their own differences" (Unger 1979, 1086). And while differences are foregrounded, it is but a small step from there to attributing difference to some biological underpinning. In turn, such explanations themselves help to generate the very differences reported in the first place. Thus, Ilan Dar-Nimrod and Steven Heine (2006), for example, reported that if women were told genetic explanations for ability in mathematics, they scored lower than if they were given social explanations of observed gender differences; hearing the biological explanation brought about difference.

A focus on difference is, however, not an accurate depiction of the natural world as a whole. Binaries pervade scientific observations of other organisms, especially in work with many vertebrates—from the genetic basis of sex determination (Fujimura 2006) to bones and health (Fausto-Sterling 2005). Yet they do not describe well all organisms: on the contrary, for many species, sexual reproduction is rare or nonexistent and "sexes" may be multiple (Fausto-Sterling 2000; Hird 2006; Wilson 2003). Even among vertebrates, anatomical sex may not be a direct consequence of chromosomes or prenatal hormones: whether the individual develops as male or female can be affected by ambient temperature, for example, while many fish species typically change sex during their lives, depending upon environmental and social conditions. And in humans, it is sometimes hard to define chromosomal sex—variations on the dichotomous pattern of XX or XY are more common than is generally assumed (Fausto-Sterling 2000). Sexual behavior, too, is not simply either/or, as some studies of animals seem to imply—as Bruce Bagemihl (1999) suggested in his review of homosexuality in the animal kingdom. In short, a simple mapping of our own insistence on binaries onto the animal kingdom, onto biological sex, or onto sexual behavior does not work.

So the first plank of feminist criticism about scientific statements of gender difference has been to reject biological determinism and its implications, to point instead to how gender is constructed and multiple. A second, related concern has been to expose the extent to which gendered biases enter the methods or interpretations of science. There is a long history of reading sexism and racism onto other species' behavior or appearance, for example, as historian of science Londa Schiebinger (1993) has described: submissive females and dominant males have abounded in the narratives of science. Ornithological guides emphasize females' "dullness" (thus, not always meriting depiction: see Van de Pitte 1998), for instance, while scientists routinely describe female fish as having "male" markings (Jackson 2001).

Researchers are now more aware of the active roles females play, across a variety of species—a shift at least partly because of feminist questions (for examples in primatology, see Fedigan 1997; Gowaty 2003). Yet assumptions about gender remain. J. Kasi Jackson (2001), for example, explains how persistently researchers see males as the norm in her field of inquiry, coloration in cichlid fish. Dominant explanations refer to male signaling and female choice; any females that show particular colorations are described as "male mimics." Such rhetoric, she points out, ignores or downplays what female behavior and markings might be doing in the worlds of these fish.

Critics identified several sources of bias: using males as standards for the species, according male behavior more status, or focusing on what males do, for example. Clearly, we can observe differences, on average, between what males do and what females do in most species of mammals or birds; we can speak of male robins being territorial, for example. But these are generalizations; they do not necessarily apply to all individuals even within a sexually dimorphic species,[6] nor do they necessarily apply to everything the animal does. And it distracts our attention from similarities and from other sources of difference.

What happens, then, when scientists write about animals and gender, is that science has *already* largely described animals within frameworks borrowed in part from our own experiencing of human society: we thus see gender differences in many other animals at least partly because we are interpreting their behavior in terms of categories already assumed—especially in those species who are most similar to us. This is not to say that there are no differences, simply that difference is exaggerated through focusing upon it. In addition to that tendency to read social constructs onto nature, there is also a tendency to generalize back again. That is, if reports appear of coy females and aggressive males in some population of nonhuman animals, then these reports will inevitably be used as models for human behavior.

Creating Gender in Scientific Studies of Animals: Through the Laboratory Lens

In this section, I move away from discussion of the general problem of reading gender onto wild nature and turn in more detail toward the practices of science in laboratories and how animals play a role in those practices. Specifically, my concern is with assumptions made when researchers look at "sex differences" and use laboratory animals to test hypotheses about such differences. Here, we can find similar assumptions about gender (or sex) in experimental reports; but what is less obvious is how these assumptions are buttressed by the very way that lab animals are kept and tested.

Nonhuman animals enter the process of producing scientific knowledge in two ways. First, scientists may study populations of specific animals in the field, in their natural environment, investigating how they fit into their habitats, how they find food, how they find mates, and so forth. Such an ecological approach follows from a long tradition of natural history, often practiced by amateur observers and only becoming professionalized in the twentieth century (Crist 1999). Or, secondly, scientists may study animals in the more constrained conditions of captivity, usually in laboratories. Here, animals may be studied as exemplars of species (investigating, for instance, how birds locate food sources within a controlled cage environment) but more often are used as exemplars of particular kinds of body plan (usually because they are mammals and are being used as "models" for human diseases). In laboratory approaches, it is not (for example) rats as such that are studied, but some sort of bodily process is investigated *using* rats. In the first case, biologists observe wild animals: they may try to control some aspect of those animals' environment (removing eggs from a clutch, for instance, to alter birds' behavior); but on the whole, natural environments are unpredictable and quite difficult to control. By contrast, in the second case, both the captive environment and the animals held captive are standardized and controlled: lab conditions are relatively predictable and animals are purpose-bred for uniformity, even becoming "tools of the trade" (Lane-Petter 1952; Rader 2004).

As I noted earlier, there is plentiful evidence of gender bias in scientific accounts of what other species of animals do and how their brains work (see also Rogers, this volume). But how do researchers' social expectations enter the controlled processes of research and knowledge production? One answer lies in how researchers make decisions about what to investigate. That is, what questions do they ask? What hypotheses? Most science is hypothesis-driven. What that means in terms of observing animals is that the human observer already has in mind a specific hypothesis, which may in principle be accepted or rejected by the observations. Even when specific hypotheses are not forthcoming (working with a species whose behavior is not well documented may, for instance, require an initial phase of descriptive observation), observers necessarily bring with them ways of seeing and coding the animals' behavior.

Choice of methods and design of experiments also affect outcome. In studying animals, this might include deciding how to categorize behavior or making decisions on specific ways of testing. Studies of sexual behavior in rodents provide a good example. Research focusing on gender differences often assumes that because there are population differences, these translate into "typical" male or female behavior. So "typical" females are said to display lordosis, a posture receptive to the male prior to coitus. This typicality can then be mapped onto some other aspect of biology—hormones for instance.

Researchers might thus look at the way that prenatal gonadal hormones can "organize" the structure and function of the brain. This concept has become well entrenched in research literature: sexual dimorphism is said to result from such early exposure to differences in hormone levels. This is called the "organization hypothesis," meaning that early hormones (prenatal or immediately postnatal) are believed to have organized the brain permanently in particular gender-specific ways. In later life, further hormonal changes can activate these behaviors.

There are certainly some ways that early hormones affect the brain, and the organization hypothesis has had significant impact on research. But the context of early development is usually ignored. The idea of organization relies on assuming typicality and ignoring other factors: for instance, in what situations is an animal's behavior "typical" of its sex? What assumptions have been made in deciding what counts as typical? Part of the context that is sometimes ignored is location in the uterus (being next to a male fetus can alter a female fetus's hormones [vom Saal 1981]) or interactions with the mother (Moore 2007). As Celia Moore (2007) emphasizes, both fetus and mother provide an environment for each other, and both affect each other. Yet the organization hypothesis predominates, such that "individual genetic differences and the importance of social interactions (even for rodents) became less visible. Hardly anyone mentioned the fact that males that had been prenatally 'organized' by testosterone still needed postnatal organization in the form of social contact," suggests Anne Fausto-Sterling (2000, 217).

The supposition of "typical" sexually dimorphic behavior based in biology has been exacerbated by the use of limited testing conditions, which do not permit animals to show their full range of behavior. Some time ago, Martha McClintock and Norman Adler (1978) pointed out how prior research on female rats had focused on their lordosis behavior—the posture permitting coitus. But this ignored the active role females play in sexual interactions. One reason for this, they argued, was the small size of the testing cages: such a barren testing arena severely restricted what the female could do, stopping her from doing most of the things (approaching, darting off, soliciting the male) that she would do given more space—behaviors which in wild animals would help to coordinate male behavior and female physiology (McClintock 1983). Accordingly, literature on mating behavior had tended to focus primarily on females as passive recipients of male advances. Biasing results through the use of particular test situations remains the case: Fausto-Sterling notes the continuing use of circular test chambers, so female rodents cannot back into corners, thus reducing unwanted "female variability" (Fausto-Sterling 2000, 221).

What research on gender difference often does is to take an average population difference (for example, that males are more likely than females to mount another animal) and then project backward into the lives of individual animals. Thus, Jill Becker and colleagues (2005), reviewing research on sex differences and hormones in humans and nonhuman animals, begin: "Female and male brains differ. Differences begin during early development due to a combination of genetic and hormonal events and continue throughout the lifespan of the individual." This claim thus elides average group differences and individual brains to *produce* an emphasis on differentiation—these "brains differ," and similarities are eroded.

What emerges, then, as scientific knowledge about gender difference, even in lab rats, is at least in part a product of how the animals are tested, in what conditions, as well as prior assumptions about what "counts" as typical behavior, appropriate to that context. When female rats do not display lordosis, for instance, they may be classified as "masculinized," while causes are sought primarily within the individual rather than the social context. It is just that removal from the social context that inspires feminist criticism—not least because feminists have long insisted that gender is something created and perpetuated in inter-individual encounters. Even for rats and mice, social context matters in generating gender.

Following Animals into the Lab

What I have sketched so far are some of the ways in which assumptions about gender difference can influence laboratory research practices and conclusions about animal behavior. Testing animals in specific, standard conditions constrains what conclusions will be drawn. But, in some ways, this problem relates to another issue—what lab animals are and how they live. The insistence in science on standardization—even of the animals themselves—contributes to the kind of context-stripping referred to above.[7] Here, I will concentrate on rodents. Although other species are, indeed, used in laboratories—many primates, dogs, birds, for example—the vast majority of animals used in lab experiments are rats and mice; and it is these that have become the most standardized. What I will argue here is that several aspects of the way that animals live and are used in labs contribute indirectly to how gender stereotypes inform research outcomes. It is not only a matter of seeing gender in animals in particular ways, but also a matter of keeping them in conditions that *create* particular kinds of knowledge. Context—or lack of it—is crucial to this argument. To explore this, we will have to try to follow the animals into the places where experiments are carried out.[8]

Yet which animals are we following? Once, scientists often used animals from a variety of sources: in the 1940s some labs were breeding rats found in waste dumps, for example (Foster 1980). Now, they are more likely to buy purpose-bred animals. Many lab animals today are created for the job (especially rodents) through breeding programs carefully designed to produce particular characteristics.

What that means in the lab is that wild animals enter here at their peril (probably to end in a mouse trap); while, in the laboratory's cages, other animals are bred carefully to meet specific experimental needs and conditions. Among other things, this might, for example, entail keeping out the kinds of diseases that their wild counterparts might bring in. So, while wild animals might live in a complex relationship with a host of other organisms, including parasites, humans, food organisms, and so forth, lab animals are removed physically from much of that context and held in cages under managed conditions. These, in turn, severely restrict social experiences and how animals react to or manage their world.

In principle, biologists study something we call "nature," but it is transformed once it enters a lab. Both the apparatuses and the animals in experimental labs are preconstructed, purpose-bred, removed from their context. As sociologist of science Karin Knorr-Cetina explains, "'Raw' materials which enter the laboratory are carefully selected and 'prepared' before they are subjected to 'scientific' tests. . . . [N]owhere in the laboratory do we find the 'nature' or 'reality' which is so crucial. . . . [T]he laboratory displays itself as a site of action from which 'nature' is as much as possible excluded rather than included" (1983, 119).

To be a lab animal is also to move from being a naturalistic animal to becoming data. Michael Lynch (1988), in a study of neuroscientists using rats, noted the contrast in how they spoke about what he termed the "analytic animal"—that is, the animal after it has become data, after its brain has been sectioned and sliced and turned into material on a slide. The phrase "that was a good animal," he observed, meant a good set of results from a well-prepared specimen. This is a very different sense of "animal" from our common sense observation of other species, he argues. Observing laboratory practices, sociologist of science Bruno Latour wryly noted that these are largely to do with "the transformation of rats and chemicals into paper." Scientists, he argues, are obsessed with graphs and diagrams but seldom see the animal who supplied the data: "Bleeding and screaming rats are quickly dispatched. What is extracted from them is a tiny set of figures" (Latour 1990, 39).

Prior to such dispatch, these "lab products" live inside the animal house rather than the lab itself. In the animal house, we find various species (though often rodents) in cages, usually in racks resembling tower blocks. The inmates may not be able to see much except their cage mates, although they can certainly smell others: the odor of a lab animal room is often overpowering. Each cage is numbered, to identify individuals.

Different rooms have different functions, breeding, surgery, record-keeping, room to house specific genetic strains; other rooms contain prepared foods, cage-washing machines. These highly structured spaces constrain, first, the behavior of the animals, who can display only limited behavior. Even if they are provided with "enrichment" in the form of toys or materials to dig, their space is very limited and quite unlike the spaces the species would inhabit in the wild. Animals are usually socially segregated—most often by sex. Second, they are also constrained in the way they are moved about within the spaces of the animal house: cages may be moved to an operating theater, into the laboratory itself (from which animals may never return), or to another room for "culling"—the routine killing of animals that have been used in experiments or because they are surplus to the laboratory's needs.

It is these spaces through which animals move between being "naturalistic" and "analytic." To staff caring for lab animals, they remain (at least partly) naturalistic; they must be cared for, protected, fed, and cleaned, and sometimes given names. They must be carefully handled, for sometimes they bite. But once they enter experimental procedures, they become the analytic animals, tools of the trade; they are given numbers and converted into data. To carry out experiments, the lab animal is often removed from its limited living space and transported elsewhere. This might entail the animal being removed from its fellows and placed for behavioral testing in a specific apparatus; if the scientist is examining something to do with sexual behavior, the animal would be placed in the testing arena with one of the opposite sex. So, the animals that enter the experiment are likely to have had rather limited life histories. Most will have had little or no experience of complex social behavior or of physical spaces outside the living cage. Some may not have had much experience of the stress of being handled by humans.

Cages and spaces (whether laboratories or testing apparatuses), then, further separate laboratories from nature. Laboratory standardization means that the movement of animals and people (researchers, technicians, veterinarians, for example) within this environment is highly constrained (Birke et al. 2007). The constraints operate according to the production line of scientific protocols, but also, ostensibly, to protect humans and animals: heavy locks on the door keep out those who are hostile to animal research while barriers protect pathogen-free strains of animals from infectious organisms.

Thus, controlling "nature" and reducing animals to numbers are part of the careful management of labs, enabling scientists to deal with variation: too much variability makes experiments difficult to interpret. Standardization—of facilities and animals—is one way that scientists have sought to minimize variation. Indeed, standardization of laboratory spaces, equipment, and living organisms historically go hand in hand (Logan 2002). In turn, standardized equipment shapes the time in which it operates and the space around which people and animals must move. Historian of science Robert Frank noted, for example, how the study of the action potential (the rapid electrical changes of a nerve impulse) depended upon development of a whole series of different devices; the animal (giant squid) supplying the nervous impulse became simply a link in a chain of equipment (Frank 1994).

Standard animals were needed to fit into the processes of increasingly regularized laboratory practices throughout the twentieth century. Rats and mice were particularly useful in this regard as they could quickly be bred for relatively uniform behavioral and physiological traits. Thus, highly specialized strains were developed (for instance, mice prone to tumors for cancer research [see Rader 2004]). In a sense, these highly inbred animals became generic, standing in as "the" laboratory model, part of the apparatus of science (Birke et al. 2007). Standardization is central to scientific experiment; it facilitates generalization. It also, importantly, constrains how animals can act, physiologically and behaviorally. But it is not, in practice, quite so easy to achieve in dealing with living animals who can react in nonstandard ways to how they are handled or tested as well as how they are housed.

One source of variability is who handles the animals. As Holmberg (2008) noted, people working in biology labs need a "feeling for the animal," they must learn how to handle their charges so as not to distress them too much.[9] In part, this requirement is supplied by courses for new researchers in animal handling, courses that place emphasis on care and ensuring good welfare. But handling remains a kind of tacit knowledge, something some people do better than others. Labs may report different results from similar experiments (Wahlsten et al. 2003; Burn et al. 2005) possibly in part because of differing personnel. In a large-scale survey of scientific data from published papers, Elissa Chesler and colleagues (2002) reported that outcomes were appreciably affected by the identity of the handler, that is, the physiological variables measured were affected by some of the factors that were controlled. In short, who does the handling matters. So, to follow an animal into the lab means tracing what interactions it has not only with physical spaces, but also with the people who simultaneously occupy that space. Some will behave toward a rat *as* an animal; others may at times treat it as simply part of the apparatus. Standardization, in short, depends on the people and practices of particular labs—a kind of local accomplishment. Indeed, what this indicates is that, far from the animal-as-data being the gold standard, the rather messier naturalistic animal pops up, an animal whose behavior and physiology are altered by its living conditions.

Perhaps unsurprisingly, animals react differently to variations in housing. Animals who live in impoverished cages differ in several ways from those who live in "enriched" ones, provided with things to do and places to go. The "naturalistic" animal, after all, has behavioral needs. Confinement tends to result in neural deficiencies, behavioral stereotypes, and anxiety (Balcombe 2006; Sherwin 2004; Würbel 2001). Particularly relevant here are reports that housing conditions can differentially affect male and female lab rats (Beck and

Luine 2002; Yildiz et al. 2007). There has, furthermore, been a longstanding tendency in research to use male animals in preference to females, with their inherent variability through the estrous cycle. How useful, then, are these animals as generalizable models—often for complex human disease conditions? How much do variations in housing or handling affect experimental outcomes? Or gender difference?

None of these points about how lab animals live their lives specifically raises issues of "gender" and its interpretation in regard to lab animals. But what they demonstrate is a science relying heavily on preconceived boundaries, not only in how scientists generate hypotheses, but also in the physical spaces of the lab and in the way that lab animals literally come to embody particular standards. The animal we might follow through the lab is an artifact, a standard, a far cry from the animals we might encounter in the outside world. Kept in confined conditions and subjected to behavioral testing in, for example, a small pen, lab animals have few choices. Furthermore, standardization removes from sight many of the messy variables (such as the people who take care of them) that would normally influence an animal's life history and reduces them to simple frameworks. Thus, if you keep animals in small groups, in tiny cages, and segregated by sex, then it might make sense to examine their sexual behavior in similar conditions.

Because science works so often through a logic of reductionism, we simply do not know how much restricted housing conditions affect gender differences in lab studies. In some ways, the results obtained in such studies are partly a product of specific testing conditions, as Fausto-Sterling (2000) noted with regard to the continuing use of circular test arenas in investigations of sexual behavior in rodents. But we know little about how relationships with people—including direct handling—affect differences between individual animals. Nor do we know to what extent standardization and selective breeding for specific traits to create "models" for disease can affect gender difference in animals.

What is known, however, is that the limited housing conditions in which lab animals are kept and how they are handled can affect their behavior and physiology. Thus, constraints on how lab animals live distort how they might behave, while constraints on how their behavior is categorized (as binary and gender-specific, for example) distort how that behavior is interpreted. If they are kept in single-sex groups, their responses and experiences are restricted long before scientists dream up any experiments. The point I want to emphasize here is that all these processes—limited housing and handling, standardized breeding, restricted test conditions—produce constraints in how we might interpret the complexity of animals' behavior. These constraints influence how we understand animals as well as how we interpret the results of experiments aimed at mapping gender difference or sexuality in other species.

While scientific research must often restrict how animals live and how they are tested in pursuit of standardization and manageable experimental design, these constraints limit the conclusions that might be drawn. And, significantly, the physical constraints of apparatuses are likely to influence observers' perceptions of animals' abilities and behavior; this, in turn, will exacerbate the focus on narrowly defined sex *difference*. What emerges as scientific knowledge about differences in lab animals is thus produced not only by how scientists think about the experiments, but also by the ways in which the testing is done—procedures that have excluded the complex interactions in which gender or sexuality might be enacted in animals living outside the lab.

Putting Animals Back into Context (and into Feminist Questions)

In following animals into the lab, my aim was to indicate how the spaces and equipment of labs constrain how people interact with animals, including through the course of experimental testing, and thus limit how the animals' behavior is seen. These constraints, in turn, can affect interpretations; thus, *even if* a researcher begins an experiment without obvious biases, gender-specific conclusions might be facilitated by the very conditions of animal housing and testing. It's a bit like funneling the complexity of an animal's behavior and possibilities into narrow containers: constraining lab animals fuels reductionist thinking, which, in turn, contributes to gendered divisions.

What I want to emphasize here, then, is that producing scientific knowledge, in part, entails assumptions scientists build into what they do—assumptions about how the natural world works—as well as their own social values. But it also substantially depends upon the *material practices* of science, both through the apparatuses of science and through the various living organisms who are compelled to enter scientific protocols. What emerges as knowledge about gender difference is thus not only a product of how scientists think, but also a product of what goes on in labs. Writing about the practices that have produced ideas of sex hormones as "messengers," Celia Roberts argues that while they "may excite or provoke sexual difference through their effects on bodies, they neither simply express nor produce sex. Hormones' messaging is received and responded to within bio-social (as opposed to purely biological) systems or worlds" (2007, 22–23). How we understand their effects is a question not only of biology but also of the various practices that produce both gender in the wider world and specific forms of knowledge from laboratory studies.

In relation to the arguments I have made here, the material practices that are implicated not only involve the specific apparatuses (see Latour 1990), but also entail the more hidden practices of animal-keeping behind the scenes. So, in the case of laboratory-based studies of gender or sexual behavior in animals, this includes both the specific testing conditions, which sometimes limit behavioral propensities in gender-significant ways, and the ways in which animals live (and die) in contrived social conditions outside of the experiment.

How, then, could feminism respond? Feminists have quite rightly pointed to various ways in which difference is perpetuated and reinforced in some experiments and reports—gender biases undoubtedly persist. But it is not enough to distinguish "sex" and "gender" only to collapse "sex" into inappropriate biological binaries; nor is it enough to point to masculinist biases. Rather, the very limitations of so many laboratory-based studies themselves hint at greater complexity in how animals develop—lab animals do *not* always respond in standard ways, despite great efforts to control variables. It is that complexity, evidenced even in the science that we might criticize, that needs to be heeded. Claims of biological gender difference are legion; acknowledgment that even lab rats have social lives shaping their development (and, hence, experimental outcomes) is much more rare. Yet it is important for both feminist theory, in particular, and the understanding of gender, more generally, to put our understanding into wider contexts. And recognizing complexity means recognizing it in the lives of animals.

Some years ago, Evelyn Fox Keller (1996) wrote about training in science as involving a process of "learning how to see"—to interpret the observed world within particular frameworks. This might entail learning to see how things are already done, such as using animals in experiments in particular ways. Or it might involve seeing animals' behavior differently, such as seeing rats as quite "adventuresome" if given the chance. Perhaps, as the opening quotations indicate, other animals "don't learn about the birds and the bees from their folks," at least as we understand such cultural lessons. They do, however, pick up subtle abilities to differentiate and communicate that do not necessarily reduce to simple neural circuits but are minimized by the very conditions in which we keep them. If only we could learn to see.

NOTES

1. I use the generic term "animal" here for convenience, although I acknowledge it is highly problematic (serving commonly to divide humans from "other animals," for example). In lab-based studies of gender or sexuality that I describe, the animals used are almost always mammals, and often rodents.
2. I refer throughout to Western technoscientific culture. Other human cultures may or may not have similar gender expectations, but my concern here is specifically with how Western science constructs and perpetuates particular ideas about gender.
3. More generally, feminist critiques of science have paid attention to how scientific knowledge is produced. That is, the knowledge and practices we call science are embedded in particular societies and always touched by those societies' beliefs and values. In that sense, "objectivity" is an ideal that ignores the specific contexts of the people who do science (Harding 1991). The role of science in producing knowledge,

then, is not to neutrally uncover truth, but is situated and contingent, always bringing with it assumptions about nature, in general, and gender, in particular.

4. As Zurbriggen and Sherman (2007) pointed out, there has been increasing use of "gender" applied to nonhuman animals in recent years. They point out that greater awareness of "gender" among scholars is desirable, but it is, they argue, problematic if authors elide sex and gender. This point is an important one; however, it rests on the assumption that other animals can be described only in terms of biology—a point that I would dispute (Birke 1994). For this reason, I persist with referring to "gender" in nonhuman animals as I do not want to collapse animal behavior back into "the biological" if the latter is taken to mean fixity. We do not know how much variability in "gendered" behavior in other species is sociocultural.
5. Now there is a journal of the same name, *Gender-Specific Medicine*. Most reports, however, are not of variables that are *specific* to one sex or other; what are reported are differences in means of populations.
6. "Sexual dimorphism" refers to the different body forms for different sexes found in some species (lions or peacocks, for example).
7. This is not intended to be a critique of scientific method as such (though it could be that); rather, my intention is to look at the interpretive consequences for how animals are housed prior to or during an experiment.
8. This section draws partly on fieldwork in labs, exploring how scientists using animals make sense of the moral dilemmas involved in doing so (see Birke et al. 2007).
9. Such empathy within our culture is stereotyped as feminine (see discussion in Birke 1994; also see Donovan 2006). These studies did not indicate to what extent animals reacted differently according to how empathic the person was, although several studies have shown that getting animals used to gentle handling reduces responses related to stress (high corticosteroid levels, for instance [see Reinhardt 2003]).

REFERENCES

Bagemihl, B. 1999. *Biological exuberance: Animal homosexuality and natural diversity*. New York: St. Martin's Press.

Balcombe, J. P. 2006. Laboratory environments and rodents' behavioural needs: A review. *Laboratory Animals* 40:217–235.

Beck, K. D., and V. N. Luine. 2002. Sex differences in behavioral and neurochemical profiles after chronic stress: Role of housing conditions. *Physiology and Behavior* 75:661–673.

Becker, J. B., A. P. Arnold, K. J. Berkley, J. D. Blaustein, L. A. Eckel, E. Hampson, J. P. Herman, S. Marts, W. Sadee, M. Steiner, J. Taylor, and E. Young. 2005. Strategies and methods for research on sex differences in brain and behavior. *Endocrinology* 146:1650–1673.

Birke, L. 1994. *Women, feminism, and animals: The naming of the shrew*. Buckingham: Open University Press.

———. 2000. Sitting on the fence: Biology, feminism, and gender-bending environments. *Women's Studies International Forum* 23:587–599.

Birke, L., A. Arluke, and M. Michael. 2007. *The sacrifice: How scientific experiments transform animals and people*. West Lafayette, IN: Purdue University Press.

Burn, C. C., A. Peters, M. J. Day, and G. J. Mason. 2005. Long-term effects of cage-cleaning frequency and bedding type on laboratory rat health, welfare, and handle-ability: A cross-laboratory study. *Laboratory Animals* 40:353–370.

Chesler, E. J., S. G. Wilson, W. R. Lariviere, S. Rodriguez-Zas, and J. S. Mogil. 2002. Influences of laboratory environment on behavior. *Nature Neuroscience* 5:1101–1102.

Crist, E., 1999. *Images of animals: Anthropomorphism and animal mind.* Philadelphia: Temple University Press.

Dar-Nimrod, I., and S. J. Heine. 2006. Exposure to scientific theories affects women's math performance. *Science* 314:435.

Donovan, J. 2006. Feminism and the treatment of animals: From care to dialogue. *Signs: Journal of Women in Culture and Society* 31:305–329.

Fausto-Sterling, A. 2000. *Sexing the body: Gender politics and the construction of sexuality.* New York: Basic Books.

———. 2005. The bare bones of sex: Part 1—Sex and gender. *Signs: Journal of Women in Culture and Society* 30:1491–1527.

Fedigan, L. M. 1997. Is primatology a feminist science? In *Women in human evolution,* ed. L. Hager, 56–75. London: Routledge.

Foster, H. L. 1980. The history of commercial production of laboratory rodents. *Laboratory Animal Science* 30:793–798.

Frank, R. G. 1994. Instruments, nerve action, and the all-or-none principle. *Osiris* 9:208–235.

Fujimura, J. H. 2006. Sex genes: A critical sociomaterial approach to the politics and molecular genetics of sex determination. *Signs: Journal of Women in Culture and Society* 32:49–82.

Gowaty, P. 2003. Sexual natures: How feminism changed evolutionary biology. *Signs: Journal of Women in Culture and Society* 28:901–921.

Haraway, D. 1989. *Primate visions: Gender, race, and nature in the world of modern science.* London: Routledge.

Harding, S. 1991. *Whose science? Whose knowledge? Thinking from women's lives.* Ithaca, NY: Cornell University Press.

Hird, M. J. 2006. Animal transsex. *Australian Feminist Studies* 21:35–50.

Holmberg, T. 2008. A feeling for the animal: On becoming an experimentalist. *Society and Animals* 16:316–335.

Jackson, J. K. 2001. Unequal partners: Rethinking gender roles in animal behavior. In *Feminist science studies: A new generation,* ed. M. Mayberry, B. Subramaniam, and L. H. Weasel, 115–119. London: Routledge.

Keller, E. F. 1996. The biological gaze. In *Future natural: Nature/science/culture,* ed. G. Robertson, M. Mash, L. Tickner, J. Bird, B. Curtis, and T. Putnam, 107–121. London: Routledge.

Knorr-Cetina, K. D. 1983. The ethnographic study of scientific work: Towards a constructivist interpretation of science. In *Science observed: Perspectives on the social studies of science,* ed. K. D. Knorr-Cetina and M. Mulkay, 115–140. London: Sage.

Lane-Petter, W. 1952. Uniformity in laboratory animals. *Laboratory Practice* (April): 30–34.

Latour, B. 1990. Drawing things together. In *Representations in scientific practice,* ed. M. Lynch and S. Woolgar, 19–68. Cambridge, MA: MIT Press.

———. 1993. *We have never been modern.* Hemel Hempstead: Harvester Wheatsheaf.

Logan, C. A. 2002. Before there were standards: The role of test animals in the production of empirical generality in physiology. *Journal of the History of Biology* 35:329–363.

Lynch, M. 1988. Sacrifice and the transformation of the animal body into a scientific object: Laboratory culture and ritual practice in the neurosciences. *Social Studies of Science* 18:265–289.

McClintock, M. K. 1983. The behavioral endocrinology of rodents: A functional analysis. *Behavioral Endocrinology* 33:573–577.

McClintock, M. K., and N. T. Adler. 1978. The role of the female during copulation in the wild and domestic Norway rat (*Rattus norvegicus*). *Behaviour* 67:67–96.

Moore, C. L. 2007. Maternal behavior, infant development, and the question of developmental resources. *Developmental Psychobiology* 49:45–53.

Nowak, M. 1980. *Eve's rib: A revolutionary new view of the female.* New York: St. Martin's Press.

Pool, R. 1994. *Eve's rib: Searching for the biological roots of sex differences.* New York: Crown Publishers.

Rader, K., 2004. *Making mice: Standardizing animals for American biomedical research, 1900–1955.* Princeton, NJ: Princeton University Press.

Reinhardt, V. 2003. Working with rather than against macaques during blood collection. *Journal of Applied Animal Welfare Science* 6:189–197.

Roberts, C. 2007. *Messengers of sex: Hormones, biomedicine, and feminism.* Cambridge: Cambridge University Press.

Schiebinger, L. 1993. *Nature's body: Gender in the making of modern science.* Cambridge, MA: Harvard University Press.

Sherwin, C. M. 2004. The influence of standard laboratory cages on rodents and the validity of data. *Animal Welfare* 13:S9–S15.

Unger, R. K. 1979. Toward a redefinition of sex and gender. *American Psychologist* 34:1085–1094.

Van de Pitte, M. M. 1998. The female is somewhat duller: The construction of the sexes in ornithological literature. *Environmental Ethics* 20:23–39.

vom Saal, F. S. 1981. Variation in phenotype due to random intrauterine positioning of male and female fetuses in rodents. *Journal of Reproduction and Fertility* 62:633–650.

Wahlsten, D., P. Metten, and J. C. Crabbe. 2003. A rating scale for wildness and ease of handling laboratory mice: Results for 21 inbred strains tested in two laboratories. *Genes, Brain and Behavior* 2:71–79.

Wilson, A. 2003. Sexing the hyena: Intraspecies readings of the female phallus. *Signs: Journal of Women in Culture and Society* 28:755–790.

Würbel, H. 2001. Ideal homes? Housing effects on rodent brain and behaviour. *Trends in Neurosciences* 24:207–211.

Yildiz, A., A. Hayirli, Z. Okumus, K. Kaynar, and F. Kisa. 2007. Physiological profile of juvenile rats: Effects of cage size and cage density. *Laboratory Animal* 36:28–38.

Zurbriggen, E. L., and A. M. Sherman. 2007. Reconsidering "sex" and "gender": Two steps forward, one step back. *Feminism and Psychology* 17:475–480.

6

"Why Do Voles Fall in Love?"

Sexual Dimorphism in Monogamy Gene Research

ANGELA WILLEY AND SARA GIORDANO

> Once upon a time, there was a meadow vole who was quite promiscuous in his behavior. He would mate with several voles and practically ignore his children. His cousin, the prairie vole, on the other hand, remained faithful to one female vole. So, scientists decided to give extra vasopressin (a hormone found in the prairie vole) receptors to the meadow voles, which have fewer vasopressin receptors. "The results were remarkable. After the V1a receptor gene was introduced, the former playboys reformed their ways. Suddenly, they fixated on one female, choosing to mate with only her—even when other females tried to tempt them," reported the BBC News.
>
> –Andrea and Alicia, *The New View on Sex*

In recent years there has been much talk in the popular press and in popular culture about monogamy and animals (e.g., Jacquet 2005). Voles have become popular in recent monogamy research because they are reported to form monogamous relationships or to be promiscuous, depending on their species. As the story above—told by chastity educators with Pregnancy Center East in Cincinnati, Ohio, on their blog, *The New View on Sex* (Andrea and Alicia 2008)—explains, prairie voles are considered a monogamous species, while the meadow vole is said not to pair-bond. Vole research on "the monogamy gene" since 2004 has received mention in the *New York Times*, *The Nation*, *Al Jazeera*, and various local papers as well as on *Late Night with David Letterman*, *The Daily Show*, *Dateline on NBC*, and National Public Radio. Some headlines focusing on genetic links to monogamy in voles read: "To Have and to Vole" (Ballon 2005), "How Geneticists Put the Romance Back into Mating" (Johnston 2005), and "Love Is a Drug for Prairie Voles to Score" (Sample 2005). A popular education module put out on YouTube by the laboratory on whose work we focus boasts a similarly provocative title: "Why Do Voles Fall in Love?"[1]

This is a play, of course, on the title of the Frankie Lymon and the Teenagers 1956, oft-covered hit song, "Why Do Fools Fall in Love?" The song made *Rolling Stone*'s top five hundred songs of all time list (November 2004); it is a deeply culturally entrenched question. The insights into the mystery that vole research on monogamy purports to offer, we argue, are similarly overdetermined. The story of "love"—in science and the larger culture of which it is a part—is clearly a gendered one. Discursively speaking, male monogamy and female monogamy both are and are not the same phenomenon to be explained. On the one hand, "monogamy" is gender neutral—it is about pair-bond formation, attaching to a mate. Both presumably heterosexual partners would need to exhibit a particular set of behaviors for a bond to be said to exist. These criteria include some form of cohabitation, mutual grooming, and co-parenting. As our introductory quote makes plain, female monogamy is more or less self-evident. It is taken for granted that a female who has mated and borne offspring is attached to her offspring's genetic other-half. We are given a fairly explicit genetic account of male mating behaviors, however, one that offers up two readings that have both been exploited in press coverage of this research and captured concisely in a cartoon from the online magazine *Science and Spirit* (Snider 2002) (see fig. 6.1). First, there is a sense of the malleability of the genetic self. Second, and implicit in the first, is the naturalization of male infidelity. This is coupled with an implicit damnation of female promiscuity, for which no genetic explanation is on offer.

Precisely because of its socially monogamous nature, "the prairie vole (*Microtus ochrogaster*) has emerged as one of the preeminent animal models for elucidating the genetic and neurobiological mechanisms governing complex social behavior" (Young Lab Web site). Because the prairie vole, unlike rats and mice, is said to be monogamous, it is fast becoming "one of the most powerful animal models for basic and translational research with direct implications for human mental health" (Young Lab Web site). In her study of the standardization of animals for experimentation in the first half of the twentieth century, Karen Rader emphasizes how cultural assumptions about animals inform their use as models in a given sociohistorical context (1995, 252). The move to using prairie voles for the modeling of human behaviors marks the consolidation of an implicit cultural consensus about monogamy as somehow fundamental to what makes the human human. We are concerned about the implications of this formulation, especially for women for whom monogamy historically *has* warranted some explanation. We are thinking here of images of black women as sexually voracious (Collins 2004), Latinas as hypersexual (Arrizón 2008), masculine lesbians as sexually predatory (Hantzis and Lehr 1994), and poor

FIGURE 6.1. "Enhanced Vasopressin Receptor" cartoon.

women, often racialized, as promiscuous mothers of too many children (Collins 1990).

As a feminist neuroscientist interested in ethics broadly construed (Sara Giordano) and a feminist theorist and historian of sexuality in science interested in monogamy's role in the production of normal and abnormal bodies (Angela Willey), we are interested in different aspects of this research and its representation in the press. Given the privileged epistemic status of science in our culture, we are both invested in understanding how assumptions about gender and difference more broadly inform its truth claims. We share a strong interest in the role that assumptions about sexual dimorphism play in the production of "a monogamy gene." Dividing populations of living things into binary sex categories for the purposes of elucidating the workings of the natural world has come to seem so logical, so obvious. Even as the efficacy of those categories is challenged within and outside of the laboratory, neuroscientific research, particularly as it relates to "love," persists in pursuing research agendas premised on the idea that physiological processes are consistent enough to warrant generalization within a sex category while being wholly different across sexes.

Our aim here is to bring literatures that have provided some insight into the production of a pre-discursive sexed body in science to bear on recent genetic research on monogamy. We are most interested in those literatures that have problematized the idea of a stable and self-evident biological sex upon which the baggage of gendered stereotypes are heaped. We argue that assumptions about binary sex are so entrenched that genetic research on monogamy continues to reproduce them in its agendas and findings, despite compelling evidence that pair bonding is not well explained in these terms. First, we contextualize understandings of sex operating in science and the culture of which it is a part at this historical moment. Second, we explain "the monogamy gene" in an effort to demystify the processes to which this language refers. Finally, we analyze the distribution across sex of research out of one of the preeminent laboratories researching monogamy in voles and offer a detailed analysis of the ways in which assumptions about the nature of sex and difference figure into the story told by neuroscientist Larry Young's laboratory about the hormone oxytocin. We draw on primary publications of the lab, supplemented by interviews Willey conducted with lab members in 2007 and 2008 to highlight continuities, contradictions, and fissures in the gendered production of a gene for monogamy.

Destabilizing Sex

The eighteenth century marked a paradigm shift in thinking about sexual difference across the study of humans, animals, and plants.[2] Rather than seeing women as slightly different (inferior) versions of men, scientists began to understand women and men as fundamentally different from one another (Schiebinger 1991, 190). The "new anatomy" supported ideas about the complimentarity of the sexes, each with their own special role. This idea, as it has developed, has played an essential role in the ongoing, though never wholly stable, sexing of the body. Contemporary neurobiological research on monogamy in animals and humans is heir to this conceptualization of sexual difference and actively incarnates it in its pursuit of explanations for differences assumed a priori.

The study of sexual strategies in general, not just monogamy, has played a role in the naturalization of male rape, female caregiving, and heterosexuality itself (Hubbard 1990). The importance of these sexual strategies hinges on the evolutionary assumption that each individual organism (human or otherwise) has as its primary "goal" the perpetuation of its own genetic material through reproduction. The differential strategies that scientists look for in women and men are linked to scientific representation of gametes: sperm are plentiful and mobile, hence males/men optimize their chances of reproduction by fucking everything. Eggs are seen as both stationary and finite in number, so females/women maximize their genetic survival by selectively choosing how to make the most of their seed. In fact, as dissenting scientific voices since at least 1948 have pointed out, this is but one way of interpreting the "lives" of the egg and the sperm. Contrary to the active/passive tale of fertilization, two cells fuse (active/active) and sperm are not as strong, goal-oriented, or potent as some scientists have imagined (Martin 1991).[3]

Assumptions about the gendered structure of pair bonding are rarely challenged as it passes unexamined as "natural." "Human nature" is "a normative concept that incarnates (in the literal sense of enveloping in flesh) historically based beliefs about how people should behave" (Hubbard 1990, 107). These beliefs are strongly steeped in ideas about gender difference and the gender division of labor. Ruth Hubbard argues that women's biology, the sexed body as it were, is political, in part because the social and biological cannot be separated:

> If a society puts half its children in dresses and skirts but warns them not to move in ways that reveal their underpants, while putting the other half in jeans and overalls and encouraging them to climb trees and play ball and other active outdoor games; if later, during adolescence, the half that has worn trousers is exhorted to "eat like a growing boy" while the half in skirts is warned to watch its weight and not get fat; if the half in jeans trots around in sneakers or boots, while the half in skirts totters about on spike heels, then these two groups of people will be biologically as well as socially different. Their muscles will be different, as will their reflexes, posture, arms, legs and feet, hand-eye coordination, spatial perception, and so on. (1990, 115)

Group differences cannot be generalized and attributed to biology in a society where different groups are treated differently and have differential access to resources. In terms of monogamy, there may well be evidence—perhaps even physiological evidence—that women desire monogamous relationships. This cannot readily be disentangled from the many compelling explanations feminists have offered for the existence of compulsory heterosexuality (Rich 1980) and compulsory monogamy (Emens 2004; Murray 1995; Rosa 1994).

The growing field of scholarship on intersexuality and the historical status of hermaphroditism in science and medicine sheds light on how the very nature of sex (and the idea that there are two) and sexuality (which relies on the two-sex model) are called into question by the existence of intersex bodies. If the sex of the biological body is made up of internal and external genitalia, hormones, chromosomes, and secondary sex characteristics (at this historical moment) and we know that there are two sexes, it would seem that the "male" and "female" versions of these traits should align themselves in any given body. They do not (Fausto-Sterling 2000). And the meanings of the ways in which bodies defy easy categorization as male and female are not clear-cut or static over time. Medical consensus about what constitutes one's "true sex" (if, indeed, one could be said to have one) has varied considerably over time. The nineteenth century was the "age of the gonads," a time where the absence or presence of testicles or ovaries held the truth of sex and any other confusing factor was interpreted as unimportant, a defect (Dreger 1998). Today chromosomes and appearance of external genitalia matter most (Kessler 1998).

At different moments (and certainly in different places), certain formations and combinations of aspects of "biological sex" are seen as essential or incidental or somewhere in between. Still, scientific research proceeds from subject recruitment and selection to the reporting of findings as if sex were binary, timeless, and self-evident. In terms of monogamy gene research, the sexing of voles is nowhere explained and the dimorphic and gendered structure of pair-bond formation is largely taken for granted. From Willey's interviews we discovered the pragmatics of vole sex assignment. The sole criterion is the distance between the anus and the dot that is their genitals. This measurement is approximate and imprecise. The only check on accuracy is that the animals are sex segregated in their cages and, if they are kept alive long enough, will sometimes reproduce, so the animals in that cage will have to be re-sexed.

With an understanding of the notion of sex as we know it as a historical process, we go on to explore how these stories about sex and sex difference show up in primary science literature. We begin with a discussion of the V1aR gene, "the monogamy gene."

A Gene for Monogamy?

In our introduction we highlighted news reports on the discovery of the monogamy gene. Although the idea of a "gene for" a particular condition or attribute is widely understood as misleading, the concept prevails in popular and scientific cultures (Hubbard and Wald 1999; Lewontin 1991). The idea of the gene preceded its materiality, that is, the discovery of DNA. The sciences and politics of eugenics were based on the concept of heritability that was fixed in the body, although a physical location of the gene did not yet exist (Hubbard and Wald 1999, Ordover 2003). The most common meaning of "gene" remains "something that is heritable." The more technical and more recent scientific definition of a gene is a region of DNA that "codes" for a specific protein. In order for a protein to be synthesized, many other proteins must be involved. Whether, how, when, and where a protein is produced, as well as how much is produced, is highly dependent on numerous factors outside the gene itself (Hubbard and Wald, 1999). Although we tend to think of genes as deterministic of behavior, there are many steps to consider between a region of DNA that has something to do with the production of a protein to how that protein interacts with other proteins and how those interactions affect a specific behavior.

The monogamy gene that is the topic of this chapter is not an on/off switch for monogamy. This is true for other genetic associations that are reported as well. For example, when being tested for a gene for Alzheimer's, you may be told you have a risk allele or not; but it is important to realize that not everyone who tests positive for that risk allele will get Alzheimer's and not everyone with Alzheimer's would test positive for that risk allele. Genetic associations are based on statistics and do not indicate a cause-and-effect relationship.

Many assumptions and simplifications must be made to make the association between behaviors understood as monogamous and the basic genetic research that is the impetus for such news reports. Many of the reports of a gene for monogamy are based on Elizabeth Hammock and Larry Young's (2005) research, in which voles were bred to have varying lengths of a specific region of DNA. The region of interest is not part of what would technically be considered a gene—it is not in the "coding" region but rather in an area surrounding the gene for a vasopressin hormone receptor, V1aR, in voles. Researchers believe that this region affects the production of these vasopressin receptors in certain brain areas. Regions of DNA outside of coding regions were until recently commonly regarded as useless stretches and often referred to as "junk DNA." Now these regions are increasingly being examined as potentially important in the regulation of protein synthesis. The voles were grouped as having either shorter or longer than average lengths of these stretches of DNA. Then these groups were compared through a series of quantifiable behavioral tests that are used as proxies for more complex behaviors such as parenting and partnering. So, for example, one result that the researchers reported was that the group with longer lengths licked their pups more often than the group with shorter lengths. These were group differences, so once again it is important to remember that if you compared a vole with a short length to one with a long length, there may be no difference in the number of times they licked their pups or the result might be the opposite of the group result.

The "monogamy gene" is not a term used in the basic scientific literatures. The phrase is a simplification—or rather an interpretation—of scientific papers used in media reports. The carefully worded title of one of the original research papers on which much news coverage was based, "Enhanced partner preference in a promiscuous species by manipulating the expression of a single gene" (Lim, Wang, et al. 2004), does not actually claim that there is a gene for monogamy. Still, the use of the phrase "single gene" evokes the idea of a gene that causes a behavior. At the same time the use of "manipulation" signals the malleability of the body and the potential of science to provide the means for changing behaviors, indeed, selves. Despite the myth that scientists are at odds with the media and annoyed by lack of public understanding, media studies have found that the relationship between scientists and the media has always been an important and beneficial one for the sciences (Lewenstein 1995, 347–348). We witnessed this in conversations with members of Young's lab. Although there was a sense that the media sometimes oversimplified their research, they track publicity on their Web site, routinely put out press releases, and have professionally framed copies of some news coverage around the lab and offices.

In news articles, we receive a familiar gendered story about monogamy, yet the monogamy gene is never qualified as the "male" monogamy gene or the monogamy gene "in males." As Bruce Lewenstein (1995) points out in his review of science and media studies, the role of science journalists has long been to "interpret" science for the public, that is, make it accessible by explaining it in simple terms. Importantly, the role of science journalists has not been to critically analyze the science or question its objectivity. So, in the press we have a gene for monogamy—an interpretation of the V1a receptor for a lay audience. What is happening in the scientific publications? Is the gendered narrative a product of the "interpretation" of science journalists, or does it run deeper?

What *is* the V1a receptor and how would regulation of the expression of such a protein affect monogamous behaviors? Much of the behavioral and genetic research on monogamy in rodents has focused on two hormones, oxytocin and vasopressin. These hormones are chemically very similar, are both released peripherally and in the brain, and are found in both males and females. The binding of these hormones to receptors in the brain is said to cause a cascade of actions that ultimately result in motivations for behaviors that are characterized as monogamous. At each stage of this process, numerous simultaneous events take place, and it is difficult to clearly define a single pathway from hormones to behavior.

Studies of vasopressin and the V1a receptor have been conducted almost exclusively in male animals. Most of the research from Young's lab on pair bonding has been on vasopressin receptors in the brains of voles. Vasopressin is widely known for its regulation of kidney function and is also used in regulation of blood pressure in humans. These peripheral effects have not been described as sex-specific. However, for its role in pair bonding or monogamy, it has been specifically investigated and linked to male behaviors. Although reports had failed to find sex-specific distribution of vasopressin receptors in both wild and lab-bred populations (Phelps and Young 2003), the search for sexual dimorphism in this system has continued. The vasopressin systems have recently been described as three separate systems where in two of the systems there is no sexual dimorphism but in the third there is some evidence of receptor distribution differences (Lim, Hammock, et al. 2004). Tracing research on the V1a receptor tells us that the pursuit of sexually dimorphic processes has shaped research on what has become "the monogamy gene" at every stage. In the next section we will examine the place of female sexuality in this research.

Sexed Research Agendas

Having problematized the idea of a "gene for" and the binary nature of "sex," we move from the theory to the laboratory and from the general to the specific. In Willey's interviews and conversations with researchers in the laboratory of neuroscientist Larry Young at Yerkes Primate Research Center, she raised questions about sex and gender frequently. They were met, generally speaking, with enthusiasm and understanding. They know better than the average non–women's studies major that sex is a fragile concept, its meaning contingent upon group consensus, sometimes as local as the lab. They know that genes are not determinants of gendered behavior. Willey experienced no great rift in our understandings of the social world and was able to explain her own academic interest in monogamy with relative ease. What she did encounter was a persistent gesture, for lack of a better word, to the results. Interviewees acknowledged that experiments are "quick and dirty" approximations of "real" processes and that, still, they work. We want to ask why they work. In this section we look at the sex-stratified and gendered experiments of the lab, based on titles and abstracts from their research publications. In the next, we zoom in on how this gendered logic reproduces itself in these experiments through assumptions about sex hormones.

Based on a Pubmed search for articles out of Young's lab that had to do with either oxytocin or vasopressin, we found sixteen primary research articles published since 2001 (see table 6.1). By reading the methods sections, we found that in five articles (31 percent) all of the subjects were female, in nine (56 percent) all subjects were male, and in two articles (13 percent) there were both male and female subjects. Four out of five of the female-only studies include "female" as a qualifier in the title of the paper. None of the nine male-only studies include a qualifier. In fact, most of these do not even specify that they are male-only in their abstracts. Female is a marked category of difference while maleness stands in as neutral. Presumably, this means that the female-only studies are only relevant for female populations and the male-only studies are more generalizable. We argue that this only makes sense from a distance. When you consider the hormones they study and the meanings attached to them, sexual difference

TABLE 6.1

Young's laboratory's publications on oxytocin and vasopressin, 2001–2009

Source/year	*Title*	*Sex of subjects*	*Hormone system studied*
Ross, Cole, et al. 2009	"Characterization of the oxytocin system regulating affiliative behavior in *female* prairie voles"	Female	Oxytocin
Ross, Freeman, et al. 2009	"Variation in oxytocin receptor density in the nucleus accumbens has differential effects on affiliative behaviors in monogamous and polygamous voles"	Female	Oxytocin
Olazabal and Young 2006a	"Oxytocin receptors in the nucleus accumbens facilitate 'spontaneous' maternal behavior in adult *female* prairie voles"	Female	Oxytocin
Olazabal and Young 2006b	"Species and individual differences in juvenile *female* alloparental care are associated with oxytocin receptor density in the striatum and the lateral septum"	Female	Oxytocin
Bielsky, Hu, Ren et al. 2005	"The V1a vasopressin receptor is necessary and sufficient for normal social recognition: A gene replacement study"	Male	Vasopressin
Bielsky, Hu and Young 2005	"Sexual dimorphism in the vasopressin system: Lack of an altered behavioral phenotype in *female* V1a receptor knockout mice"	Female	Vasopressin
Hammock et al. 2005	"Association of vasopressin 1a receptor levels with a regulatory microsatellite and behavior"	Male	Vasopressin
Hammock and Young 2005	"Microsatellite instability generates diversity in brain and sociobehavioral traits"	Male	Vasopressin

(*continued*)

Table 6.1. Young's laboratory's publications on oxytocin and vasopressin, 2001–2009 (*continued*)

Source/year	*Title*	*Sex of subjects*	*Hormone system studied*
Nair et al. 2005	"Central oxytocin, vasopressin, and corticotropin-releasing factor receptor densities in the basal forebrain predict isolation potentiated startle in rats"	Male	Oxytocin and Vasopressin
Bielsky et al. 2004	"Profound impairment in social recognition and reduction in anxiety-like behavior in vasopressin V1a receptor knockout mice"	Male	Vasopressin
Lim, Murphy, et al. 2004	"Ventral striatopallidal oxytocin and vasopressin V1a receptors in the monogamous prairie vole (*Microtus ochrogaster*)"	Both	Vasopressin
Lim, Wang, et al. 2004	"Enhanced partner preference in a promiscuous species by manipulating the expression of a single gene"	Male	Vasopressin
Lim and Young 2004	"Vasopressin-dependent neural circuits underlying pair bond formation in the monogamous prairie vole"	Male	Vasopressin
Phelps and Young 2003	"Extraordinary diversity in vasopressin (V1a) receptor distributions among wild prairie voles (*Microtus ochrogaster*): Patterns of variation and covariation"	Both	Vasopressin
Ferguson et al. 2001	"Oxytocin in the medial amygdala is essential for social recognition in the mouse"	Male	Oxytocin
Pitkow et al. 2001	"Facilitation of affiliation and pair-bond formation by vasopressin receptor gene transfer into the ventral forebrain of a monogamous vole"	Male	Vasopressin

emerges as a pursuit of monogamy gene research. Almost all of the studies on oxytocin were conducted in female-only populations, and almost all of the studies on vasopressin were conducted in male-only populations. Why is this and (why) does it "work"?

Is Oxytocin from Venus?

According to researchers, the release of certain hormones during mating, between a male and female, causes the animals to bond—if they are monogamous. In females, the hormone said to facilitate pair bonding is oxytocin, which is also widely understood to control maternal care behaviors. Pair bonding in males is said to be facilitated by the hormone vasopressin. Vasopressin is not linked specifically to bonding or caregiving behaviors, but rather is said to control "species-specific" behaviors in males. As neuroscientist Larry Young explained in an interview, vasopressin makes hamsters territorial; it makes prairie voles bond with a female and want to protect her and their young. This research agenda reflects a story about monogamy that is strangely familiar—females become intensely attached to whoever they have sexual intercourse with and males do not (Dallos and Dallos 1997, 139–141). When males settle down, they are motivated by feelings of possessiveness.[4] We offer analyses of two sets of assumptions informing this gendered story: (1) definitions of sex (the act) that serve as the rationale for oxytocin's centrality to pair bonding and (2) the notion of "the maternal brain." These assumptions are so deep and so persistent that evidence disrupting them tends to be ignored or compartmentalized, such that research on these "monogamy" hormones proceeds with veracity along sexed lines.

Heather Ross and Larry Young's account of how oxytocin became a prime candidate for pair-bonding research is telling: "Because of the role that [oxytocin] plays in mother-infant attachment and since mating results in vaginocervical stimulation, which is known to release oxytocin in the brain, oxytocin was a prime candidate for regulating the formation of the pair bond" (2009, 538). Two foundational assumptions are operating here: (1) mating is somehow causally linked to pair-bond formation and (2) mother-infant bonding is either similar to or linked with pair bonding. We want to question the obviousness of these implicit claims.

"Mating" causes a pair bond to form, and mating is coded as vaginal intercourse. For females "vaginocervical stimulation" is the behavioral mechanism that starts this process. Although it is not discussed to nearly the extent the cervix is, it would seem that orgasm is what triggers this neuro-chemical process for males. The existence of a potentially analogous phenomenon in females leads Young (2009), in a recent essay in the opinions section of *Nature,* to laud the evolutionary wonder of cervical stimulation as part of sexual intimacy.

The lack of widespread consensus about the joys of cervical exams or childbirth have led feminists to question evolutionary explanations of female sexuality that link it to, or rather explain it, in terms of reproductive sex.[5] Elisabeth Lloyd (1993) calls this "the orgasm-intercourse discrepancy." Specifically, she interrogates ideas about orgasm as an evolutionary adaptation that rewards female primates for having frequent sex with their male mates. The linking of orgasm to intercourse, she argues, is a fairly ludicrous yet totally unquestioned assumption underlying the formulation of questions and experiments across a broad range of research on female sexuality. A version of this "discrepancy" seems to be foundational to our understanding of the naturalness of female monogamy here.

In fact, a great many things trigger the release of oxytocin and could be understood to facilitate the formation of bonds between animals. Small amounts of oxytocin are released over time when the animals spend long periods of time together; nipple stimulation, touching, and "grooming" cause oxytocin to be released; and hormones released during exercise may also bind with these same receptors.

Even as cervical stimulation is sometimes acknowledged as at least potentially inessential to pair bonding (Pitkow et al. 2001; Ross and Young 2009), oxytocin is persistently feminized. Oxytocin (from Greek, "rapid birth") was first named because it was found to play a role in facilitating the birthing of offspring. More recently, and related to the themes of the current research on monogamy, oxytocin has been reported to play a role in the brain by motivating "maternal behaviors"—those which are "necessary for the survival of the offspring" (Ross and Young 2009). Throughout the scientific literature, oxytocin is so tightly linked to maternal behaviors that evidence suggesting that oxytocin is important in males or is not as important in females is almost unintelligible. Somehow, the rapid-birth hormone cannot be interpreted as anything other than an essential female/maternal hormone. In the same review article in which the authors note that, although "early studies have focused on the role of oxytocin in regulating female behavior with a focus on reproduction," it is important in regulating many behaviors in males and females, the authors also dedicate a significant portion of the paper to how oxytocin regulates the "maternal brain" (Ross and Young 2009, 534).

Anne Fausto-Sterling (2000) provides a detailed analysis of the history of the so-called sex hormones, estrogen and testosterone. She shows how culturally influenced the sexing of their functions is. Even in the naming of the hormones, she points out, the female hormones are linked to reproduction by being named for the estrus cycle while the male hormones are named more broadly as the hormone that makes one male. Like Fausto-Sterling, we see our cultural assumptions about sex differences and the role of motherhood shaping research and discussion of oxytocin and vasopressin. For example, the use of the term "maternal brain" is essentializing, suggesting both that all women are mothers and that there is a universal experience of motherhood. In contrast, we do not find discussions of the "paternal brain" in any of the literature. While oxytocin is deeply associated with maternal behaviors across species, vasopressin research in male rodents describes its role in species-specific male behaviors, with parenting claimed as one of these in voles.

Oxytocin continues to be understood as the cause of universal cross-species maternal behavior, even as conflicting evidence emerges within the lab. In 2001, Thomas Insel and colleagues reviewed evidence that surprised researchers in an article titled "Oxytocin: Who Needs It?" The authors concluded "not mice" based on a study that showed that mice with a null mutation for oxytocin "exhibited full maternal and reproductive behaviors" (Insel et al. 2001, 59). They went on to suggest that primates may not need oxytocin for these behaviors either. These data, even as they are cited in future research papers, become inassimilable in the retelling of the story of oxytocin's role in maternal care: the narrative is consistently universalized and persistently cited as evidence of the efficacy of putting funding dollars into research on the hormone.

Despite highly sex-specific research on oxytocin, the hormone and receptors are found in both males and females. It has been reported that there are no sex differences in the distribution of oxytocin receptors in the brains of voles (Lim, Murphy, et al. 2004). The conclusion they draw is that some other part of the oxytocin system must be sexually dimorphic. So we begin with two assumptions: that males and females are different and that oxytocin is a female/maternal hormone. Based on these pre-theoretical truths, the search for sex differences and oxytocin's role in manifesting them is a central and ongoing tenet of vole research on monogamy.

A Persistent Narrative

Over the course of our research, more and more clinical trials were conducted based on Young's lab's findings. In humans, these usually involved the administration of oxytocin—in both males and females. Willey asked each of her three interviewees to explain this discrepancy: "We portray it as being exclusive in the voles that males use vasopressin, and females oxytocin. But there is actually evidence that there is overlap." Interviewee B goes on to explain that there is actually a lot of "cross-receptivity" in receptors, meaning that it is not always clear which hormone is causing the behavior. Interviewees A and C both acknowledged that both males and females have both hormones (oxytocin and vasopressin) and suggested that the role of oxytocin in male bonding is likely underestimated.

Still, when the story is told in scientific journals, Young gives a much neater description of sexual dimorphism: "Pair bonding in males involves similar brain circuitry to that in females, but different neurochemical pathways. In male prairie voles, for example, vasopressin—a hormone related to oxytocin—stimulates pair bonding, aggression towards potential rivals, and paternal instincts, such as grooming offspring in the nest" (Young 2009).

Lewenstein's review of science and media studies highlights studies of television programming on science in which the acknowledgment of "uncertain outcomes end[s] up as displays of certainty" (1995, 355). In both the published literature from the Young lab and interviews, this phenomenon seems to be at work—caveats about uncertainty or possible alternate outcomes actually help legitimate the story they tell. As Anne Fausto-Sterling describes, although the answers based on scientific research are underdetermined (that is, a result could be explained by multiple "plausible" explanations), we must consider that there are many factors—including cultural assumptions about sexual difference—that make one answer stand out as a better one as scientific truths are disseminated (2000, 162).

Is it possible that vasopressin and oxytocin are not so gendered as we imagine? What could we learn to know about human attachment, about "love," if we could imagine it outside of historically situated ideas about gender and heterosexuality? If the naturalness of maternal care, the link between love and reproduction, and the primacy of sexual difference were questions, not the stable ground from which we do the asking, what would we want to know?

ACKNOWLEDGMENTS

Many thanks to Nancy Campbell, Sander Gilman, Lynne Huffer, Mark Jordan, Deboleena Roy, Pamela Scully, and the Young laboratory and to the community and fellowship support of the Bill and Carol Fox Center for Humanistic Inquiry at Emory University.

NOTES

1. The video can be viewed at http://www.youtube.com/watch?v=Oh8x9KDkYTc (accessed November 21, 2010).
2. Though most plants' flowers possess both stamen and pistil (which are now considered to be the male and female parts, respectively), investment in sexual dimorphism had a powerful influence on how plant reproduction was conceptualized. The seed-producing plants (or parts of plants) were automatically assumed to be female (because the seed was thought of as an egg), while much debate incurred over what the important "male" part was. Not only was sexual difference the implicit framework by which reproduction was understood (and thus knowledges about so many aspects of the natural world explained), but explicitly anthropomorphic metaphors were used to describe and explain plant reproduction. Linnaeus's plants married, made love, and had conflict.
3. See also Irigaray's *Speculum of the Other Woman* (1985, 15) for another approach to deconstructing the logic of sperm and egg as active and passive.

4. This naturalization of male jealousy takes many forms, is well documented, and has actually informed court cases wherein men have committed "crimes of passion." See, for example, Daly, Wilson, and Weghorst's "Male Sexual Jealousy" (1982).
5. See Anne Koedt's *The Myth of the Vaginal Orgasm* (1970).

REFERENCES

Andrea and Alicia. 2008. Oxytocin, vasopressin, and a tale of two voles. *The New View on Sex*, April 4. http://thenewviewonsex.blogspot.com/2008/04/oxytocin-vasopressin-and-tale-of-two.html (accessed September 30, 2009).

Arrizón, A. 2008. Latina subjectivity, sexuality and sensuality. *Women and Performance: A journal of feminist theory* 18 (3): 189–198.

Ballon, M. S. 2005. To have and to vole. *Philippine Daily Inquirer*, June 18, 4.

Bielsky, I. F., S. B. Hu, X. Ren, E. F. Terwilliger, and L. J. Young. 2005. The V1a vasopressin receptor is necessary and sufficient for normal social recognition: A gene replacement study. *Neuron* 47 (4): 503–513.

Bielsky, I. F., S. B. Hu, K. L. Szegda, H. Westphal, and L. J. Young. 2004. Profound impairment in social recognition and reduction in anxiety-like behavior in vasopressin V1a receptor knockout mice. *Neuropsychopharmacology* 29 (3): 483–493.

Bielsky, I. F., S. B. Hu, and L. J. Young. 2005. Sexual dimorphism in the vasopressin system: Lack of an altered behavioral phenotype in female V1a receptor knockout mice. *Behavioural Brain Research* 164 (1): 132–136.

Collins, P. H. 1990. *Black feminist thought: Knowledge, consciousness, and the politics of empowerment.* London: Unwin Hyman.

———. 2004. *Black sexual politics: African Americans, gender, and the new racism.* New York: Routledge.

Dallos, S., and R. Dallos. 1997. *Couples, sex, and power: The politics of desire.* Philadelphia: Open University Press.

Daly, M., M. Wilson, and S. Weghorst. 1982. Male sexual jealousy. *Ethology and Sociobiology* 3 (1): 11–27.

Dreger, A. D. 1998. *Hermaphrodites and the medical invention of sex.* Cambridge, MA: Harvard University Press.

Emens, E. F. 2004. Monogamy's law: Compulsory monogamy and polyamorous existence. *New York University Review of Law and Social Change* 29:277.

Fausto-Sterling, A. 2000. *Sexing the body: Gender politics and the construction of sexuality.* New York: Basic Books.

Ferguson, J. N., J. M. Aldag, T. R. Insel, and L. J. Young. 2001. Oxytocin in the medial amygdala is essential for social recognition in the mouse. *Journal of Neuroscience* 21 (20): 8278–8285.

Hammock, E. A., M. M. Lim, H. P. Nair, and L. J. Young. 2005. Association of vasopressin 1a receptor levels with a regulatory microsatellite and behavior. *Genes Brain and Behavior* 4 (5): 289–301.

Hammock, E. A., and L. J. Young. 2005. Microsatellite instability generates diversity in brain and sociobehavioral traits. *Science* 308 (5728): 1630–1634.

Hantzis, D. M., and M. V. Lehr. 1994. Which desire? Lesbian (non) sexuality and TV's perpetuation of hero/sexism. In *Queer words, queer images: Communication and construction of homosexuality*, ed. R. J. Ringer, 107–121. New York: New York University Press.

Hubbard, R. 1990. *The politics of women's biology.* New Brunswick, NJ: Rutgers University Press.

Hubbard, R., and E. Wald. 1999. *Exploding the gene myth: How genetic information is produced and manipulated by scientists, physicians, employers, insurance companies, educators, and law enforcers.* Boston: Beacon Press.

Insel, T. R., B. S. Gingrich, et al. 2001. Oxytocin: Who needs it? *Progress in Brain Research* 133:59–66.

Irigaray, L. 1985. *Speculum of the other woman.* Ithaca, NY: Cornell University Press.

Jacquet, L. 2005. *March of the Penguins.* Warner Independent Pictures.

Johnston, I. 2005. How geneticists put the romance back into mating. *The Scotsman,* July 30, 34.

Kessler, S. J. 1998. *Lessons from the intersexed.* New Brunswick, NJ: Rutgers University Press.

Koedt, A. 1970. *The myth of the vaginal orgasm.* AgitProp Literature Programme. See online at http://www.uic.edu/orgs/cwluherstory/CWLUArchives/vaginalmyth.html.

Lewenstein, B. V. 1995. Science and the media. In *Handbook of science and technology studies,* ed. S. Jasanoff, G. E. Markle, J. C. Petersen, and T. Pinch, 343–360. Thousand Oaks, CA: Sage.

Lewontin, R. C. 1991. *Biology as ideology: The doctrine of DNA.* Toronto: House of Anansi.

Lim, M. M., E. A. Hammock, and L. J. Young. 2004. The role of vasopressin in the genetic and neural regulation of monogamy. *Journal of Neuroendocrinology* 16 (4): 325–332.

Lim, M. M., A. Z. Murphy, and L. J. Young. 2004. Ventral striatopallidal oxytocin and vasopressin V1a receptors in the monogamous prairie vole (*Microtus ochrogaster*). *Journal of Comparative Neurology* 468 (4): 555–570.

Lim, M. M., Z. Wang, D. E. Olazábal, X. Ren, E. F. Terwilliger, and L. J. Young. 2004. Enhanced partner preference in a promiscuous species by manipulating the expression of a single gene. *Nature* 429 (6993): 754–757.

Lim, M. M., and L. J. Young. 2004. Vasopressin-dependent neural circuits underlying pair bond formation in the monogamous prairie vole. *Neuroscience* 125 (1): 35–45.

Lloyd, E. A. 1993. Pre-theoretical assumptions in evolutionary explanations of female sexuality. *Philosophical Studies* 69 (2): 139–153.

Martin, E. 1991. The egg and the sperm: How science has constructed a romance based on stereotypical male-female roles. *Signs 16* (3): 485–501.

Murray, A. 1995. Forsaking all others: A bifeminist discussion of compulsory monogamy. In *Bisexual politics: Theories, queries, and visions,* ed. N. Tucker. New York: Hayworth Press.

Nair, H. P., A. R. Gutman, M. Davis, and L. J. Young. 2005. Central oxytocin, vasopressin, and corticotropin-releasing factor receptor densities in the basal forebrain predict isolation potentiated startle in rats. *Journal of Neuroscience* 25 (49): 11479–11488.

Olazabal, D. E., and L. J. Young. 2006a. Oxytocin receptors in the nucleus accumbens facilitate "spontaneous" maternal behavior in adult female prairie voles. *Neuroscience* 141 (2): 559–568.

———. 2006b. Species and individual differences in juvenile female alloparental care are associated with oxytocin receptor density in the striatum and the lateral septum. *Hormones and Behavior* 49 (5): 681–687.

Ordover, N. 2003. *American eugenics: Race, queer anatomy, and the science of nationalism.* Minneapolis: University of Minnesota Press.

Phelps, S. M., and L. J. Young. 2003. Extraordinary diversity in vasopressin (V1a) receptor distributions among wild prairie voles (*Microtus ochrogaster*): Patterns of variation and covariation. *Journal of Comparative Neurology* 466 (4): 564–576.

Pitkow, L. J., C. A. Sharer, X. Ren, T. R. Insel, E. F. Terwilleger, and L. J. Young. 2001. Facilitation of affiliation and pair-bond formation by vasopressin receptor gene transfer into the ventral forebrain of a monogamous vole. *Journal of Neuroscience* 21 (18): 7392–7396.

Rader, K. A. 1995. *Making mice: Standardizing animals for American biomedical research.* Bloomington: Indiana University.

Rich, A. 1980. Compulsory heterosexuality and lesbian existence. *Signs* 5 (4): 631–660.

Rosa, B. 1994. Anti-monogamy: A radical challenge to compulsory heterosexuality. In *Stirring It: Challenges for Feminism*, ed. G. Griffin, M. Hester, and S. Rai, 107–120. London: Taylor and Francis.

Ross, H. E., C. D. Cole, Y. Smith, I. D. Neumann, R. Landgraf, A. Z. Murphy, and L. J. Young. 2009. Characterization of the oxytocin system regulating affiliative behavior in female prairie voles. *Neuroscience* 162 (4): 892–903.

Ross, H. E., S. M. Freeman, L. L. Spiegel, X. Ren, E. F. Terwilliger, L. J. Young. 2009. Variation in oxytocin receptor density in the nucleus accumbens has differential effects on affiliative behaviors in monogamous and polygamous voles. *Journal of Neuroscience* 29 (5): 1312–1318.

Ross, H. E., and L. J. Young. 2009. Oxytocin and the neural mechanisms regulating social cognition and affiliative behavior. *Frontiers in Neuroendocrinology* 30 (4): 534–547.

Sample, I. 2005. Love is a drug for prairie voles to score. London *Guardian*, December 5, 5.

Schiebinger, L. 1991. *The mind has no sex? Women in the origins of modern science.* Cambridge, MA: Harvard University Press.

Snider, C. J. 2002. What's love got to do with it? *Science and Spirit Magazine.* Cartoon. www.science-spirit.org (accessed September 30, 2009).

Young, L. J. 2009. Being human: Love: Neuroscience reveals all. *Nature* 457 (7226): 148.

Young Lab. Vole Genomics Initiative. Emory University. http://research.yerkes.emory.edu/Young/volegenome.html (accessed September 30, 2009).

7

What Made Those Penguins Gay?

Gender and Sexuality Politics in the Zoo

K. SMILLA EBELING AND BONNIE B. SPANIER

> As a matter of course we will accept the male couples as they are, and we certainly won't coercively heterosexualize them, as we have been accused of for the last year.
>
> –Heidi Kück, "Pressemitteilung Zoo am Meer Bremerhaven"

Introduction: Science, Zoos, and Gay Families

The opening epigraph refers to penguins, more precisely, to male penguins that show same-sex sexual behavior and the ability to procreate and to build a family.[1] Why might the sexual activities of penguins be important for people of Western industrialized societies? Why do some U.S. parents not want their children to read a children's book about gay penguins? And for what reason do gay and lesbian communities fight for gay penguin couples in a German zoo to stop the zoo from encouraging them to mate with females? Our chapter explores the following key issues: homosexuality, families, hetero- and homonormativity, and the understanding of human animals in relation to nonhuman animals and vice versa.

Scientific knowledge enjoys high public confidence and functions as a powerful authority in society. People generally regard science as value-neutral and impartial. Hence, scientific claims, in our case about the sex and sexuality of nonhuman animals, are regarded as truth descriptions. In heteronormative-structured thinking, reproductive opposite-sex sexuality has found easy explanations in biology, but biological explanations—causes and functions—of homosexual behavior, whether by human or nonhuman animals, has created controversy (e.g., see Spanier 2000). One crucial point of explanation and legitimization concerns whether certain sexual variations exist in "nature"—meaning in nonhuman animals. Some scientists claim, for example, that nonhuman animals do not show same-sex sexual behavior and, thereby, claim that homosexuality in human animals is "unnatural." Others claim that some nonhuman animals are homosexual, but that this is a rare error of nature. A different perspective recognizes homosexuality as occurring among nonhuman animals and, therefore, as "natural." Some scientists extrapolate, explicitly or implicitly, from nonhuman animal behavior studies in and beyond the laboratory to claim that sexual orientation among human animals is rooted in biology, which may refer to anatomical or genetic differences that are linked to evolution.

Various scholars of feminist science studies have long pointed to the busy traffic in both directions between nature/nonhuman animals and culture/human animals, especially when it comes to biological explanations of social practices and human animal behaviors (e.g., Ebeling 2002; Fausto-Sterling 2000; Gowaty 1981; Haraway 1989; Schiebinger 1993; Spanier 1995). Following these insights, knowledge production about nonhuman animals occurs in the frame of heteronormative thinking, so it is not surprising that the lens of zoological knowledge about sex, gender, and sexuality fosters the idea that sexual dimorphism, dioecy, and opposite-sex sexual behavior are the prevalent and, thus, "normal" forms of sex, gender, and sexuality found in nature. Only a relatively few zoological studies have examined homosexuality in nonhuman animals, despite documentation within several studies. Moreover, there was no major effort (meaning no big research projects funded by high-prestige science foundations that do systematic research in all animal groups) to study homosexual nonhuman animals in zoology until the end of the twentieth century. Recently, however, two remarkable scientific books on homosexuality and transgender in nonhuman animals, entitled *Biological Exuberance* (Bagemihl 1999) and *Evolution's Rainbow* (Roughgarden 2004), provided detailed knowledge of same-sex sexuality and transgender behavior in nonhuman animals, describing widespread and diverse homosexualities in nature, reflecting on the scientific community's treatment of it, and suggesting non-heteronormative theories of sexuality that do not focus on reproduction.

Public debates on sex and sexuality in human animals frequently refer to zoological descriptions of nonhuman animal sexual behavior and use them to make sexualities legitimate (or not) by considering them as natural (or unnatural) and, by a common logic, acceptable (or not). This reasoning has been used as a powerful argument against anything other than heterosexuality in society. Since people look at nonhuman animals as role models—perhaps because people cannot help but relate nonhuman animal behavior to human animal behavior (Zuk 2002) or because people like to do it (Haraway 1989) or because it is a scientific method (Terry 2000)—we regard knowledge about sex and sexuality in the animal kingdom as crucial for society. For a few years now, some public institutions like zoos and natural history museums have begun to educate people about the variety of sexes and sexual behaviors in nonhuman animals. For instance, the zoos in Basel (Switzerland), Zürich (Switzerland), and Amsterdam (the Netherlands) offer thematic guides on homosexuality in nonhuman animals, for which *Biological Exuberance* provided much of the scientific knowledge. And the London Zoological Society organizes parties called "Gay Sunday" in the London Zoo in Regents Park in order to encourage a debate on homosexuality in nonhuman animals. Moreover, the Museum of Natural History in Oslo, Norway, showed the successful exhibition "Against Nature? An Exhibition on Animal Homosexuality" in 2006–2007, which drew crowds to the Oslo zoo and is now available as an international touring exhibition (Naturhistorisk Museum n.d.).

Those public events were made possible by societal changes. According to the guiding biologist in the Basel Zoo, it was not the zoo's idea to start the homosexuality guide, but gay and lesbian communities asked for it in the context of "Pink May," a month in which Basel celebrates homosexuality through a variety of cultural events. In response, the zoos had to educate themselves about nonreproductive and other sexual behaviors that do not fit a heteronormative paradigm. Thus, they had to find biological knowledge about homo-, trans-, and intersexuality in nonhuman animals. Supported by gay celebrities, like Sir Elton John, Dale Winton, and Graham Norton, the idea for the "Gay Sunday" in London's zoo was triggered by the insight that the zoo had become a top venue for gay dating (Knowles 2006). One of the reasons the Museum of Natural History in Oslo mounted the exhibition "Against Nature?" was to use a provocative issue to attract a large number of visitors (Beyerstein 2006; personal communication with museum director). Those events in public institutions illustrate the importance of public concern emerging from gay and lesbian communities and affecting educational displays about nonhuman animals. Here we see politics (changed life practices and lived sexualities) and science (changed understandings of the sexes and sexualities in society) intertwined.

The issue of homosexuality in human and nonhuman animals involves the question of how society understands people in relation to nonhuman animals and vice versa. The theoretical basis of our (the authors') understanding of this human-animal relation is the approach of "thinking with animals," which applies to science as much as to culture. In the words of Lorraine Daston and Gregg Mitman, thinking with nonhuman animals is "the irresistible taboo [of assuming] that animals are like us" and that "humans assume a community of thought and feeling between themselves and a surprisingly wide array of animals; they also recruit animals to symbolize, dramatize, and illuminate aspects of their own experience and fantasies" (2005, 1, 2).[2] Based on this view, we assume that people use nonhuman animals in order to think about human animals and society, that people express themselves through nonhuman animals, and that people believe they know how nonhuman animals think and feel. This occurs, for example, when a zookeeper says that a nonhuman animal "desperately wants to hatch an egg" or pet owners think that their dog is happy to see them when they come home. People also commonly express their feelings and thoughts by translating nonhuman animals into symbols, for instance, by using nonhuman animals in swearwords or pet names or by depicting nonhuman animals as messengers in greeting cards. Hence, people are used to superimposing human animal sense meanings onto nonhuman animals and then drawing conclusions about the human animal world by observing nonhuman animals, which is a form of circular reasoning. Here we apply this understanding of thinking with nonhuman animals to illustrate that knowledge about gender and sexuality in human animals interplays with (or co-constructs) knowledge about nonhuman animal sexual behaviors, as can be seen in public debates about gay nonhuman animals. Our analysis also points to the circular ways in which descriptions of nonhuman animal sexual behaviors influence cultural ideas about the sexes and sexualities in human society; these ideas, in turn, are used to legitimatize gender relations in society.

In order to scrutinize how thinking with nonhuman animals affects the production of public meaning about sex and sexuality, we focus on zoos as public educational institutions. Zoos are important places for urban people to meet and appreciate actual nonhuman animals (Berger 1980), with the number of visitors to zoos topping the number attending baseball, hockey, and football games (Breadsworth and Bryman 2001). For the public, zoos are a highly significant source of information about nonhuman animal lives. With increasing numbers of zoo-soaps on U.S. and German television, zoos get even more public attention, amplifying the significance of zoos as mediators of knowledge about nonhuman animals.[3]

Within zoos, our focus is on penguins because the last few years have witnessed the eruption of public debates about gay penguins in German and U.S. zoos. We will first analyze recent attention, from 2005 to 2007, about six gay male penguins in a German zoo and then an earlier fuss about two male penguins in New York's Central Park Zoo that lived for six years as a gay couple and then broke up in 2004. Secondly, we will examine two children's books, written independently by American and German authors, about the penguin pup "Tango," who was raised by the two Central Park Zoo penguins.

Penguins have become prominent nonhuman animals in pop culture over the last five years, as the feature-length movies *March of the Penguins* (2006), *Happy Feet* (2007), and *Surf's Up* (2007) show.[4] Penguins might be popular partly because they are cute, being bipedal and looking as if they are wearing formal tailcoats, and perhaps because they walk in a clumsy way, which is amusing to watch. Those movies grabbed the public's imagination about penguins and served as a focal point to support traditional heterosexual family and antigay politics.[5] Working with penguin stories in popular culture, we will describe how thinking with nonhuman animals helped negotiate human animal sexualities. We will also explore what we can learn from thinking with penguins.

Gay Penguins in German and U.S. Zoos

Everybody Is Free to Pursue His/Her Own Happiness: Gay Penguins in Bremerhaven's Zoo

In spring 2005 and 2006, six male Humboldt penguins that had coupled in three same-sex pairs in the Zoo am Meer Bremerhaven (Bremerhaven, Germany) received a great deal of attention in newspapers across the political spectrum from conservative to left wing. The news coverage in those two years reflected the same spectrum with regard to gay issues in society.

The Zoo am Meer Bremerhaven wanted their gay male penguins to reproduce, with the aim of preventing the endangered Humboldt penguins from becoming extinct. The zoo's director assumed that the males coupled in same-sex pairs due to a lack of females (a common claim about homosexual behaviors in nonhuman animals [see Driscoll 2008]), so she imported female penguins from Sweden. Claiming to test the penguins' homosexuality, the director asserted the zoo had a scientific goal, since experts had observed that homosexual pairs in some species could not be separated easily. The zoo director was expressing typical zoo narratives of conducting research and protecting species from extinction, both of which are important tasks of today's zoos that legitimize the capture and exhibition of wild nonhuman animals.

The news coverage performed its business accordingly. Germans tend to see Swedish women as highly feminine, attractive, sexy, and very open-minded when it comes to engaging in sex. Sex sells, and sex stories about nonhuman animals sell too, so it is no wonder that all the newspapers used language associating the Swedish penguin imports with Swedish women. They highlighted, for instance, that the Swedish "ladies" would seduce or convert the gay penguins and used the title "Swedish allurement" (Deutsche Presse Argentur, *Sueddeutsche* 2005; Deutsche Presse Argentur, *Taz* 2005). But the press reported, "Converting attempt fails" and the "gay penguins stand firm" (Associated Press, *Sueddeutsche* 2005; Deutsche Presse Argentur, *Stern* 2005). The male penguins did not show any interest in the females and coupled in both years in same-sex pairs. The media used human animal concepts such as true love and homosexual affection to describe penguins' behavior. In 2005, the director assumed that the female penguins probably had arrived too late, that is, after the male couples had already built a strong bond. In 2006, she explained the "true love only among men" with the observation that the males abstained from "the ladies," yet, despite the evidence, repeated her understanding that "men [*sic*] pair-bonding is not a question of self-realization but nature's fancy caused by a lack of females" (Deutsche Presse Argentur, *Stern* 2006).

Gay and lesbian communities around the world contributed to the media's provocative and controversial nonhuman animal sex stories by protesting against the zoo's treatment of the gay penguins. The news coverage headlined "Coercive harassment through female persuasive talent," "Gay penguins stirred emotions worldwide," and "Homosexuals fight for gay penguins" (Associated Press, *Stern* 2005; Heume 2005; Deutsche Presse Argentur, *Ärzte Zeitung* 2005). The zoo received obscenities by phone calls and e-mails as well as the attention of radio and TV stations from Austria, Australia, the United States, and Germany. Terms used in the protests were "juridical minor wards," "comparability with homosexual cohabitations," "coercive conversion," and "right of unswayed coupling." The zoo reacted to this bad press by stating that nobody wanted to reeducate or heterosexualize the gay penguins. "Here, everybody is free to pursue his/her own happiness"; the gay penguins are allowed to stay gay and nobody will encroach on their sexual self-determination (Associated Press, *Stern* 2005). "If they are gay, then they are gay" (Kiehne 2005). In the second year of experimenting with homosexual male penguins, the zoo eagerly explained, "We take it for granted that we accept the male pairs as they bond, and we do not heterosexualize them compulsorily" (Kück 2006).[6]

Why are nonhuman animal sex stories like these penguin stories of great interest to the public? Following the approach of thinking with nonhuman animals, people do more than write and think about gay penguins when they read or write about them. They naturalize cultural values and ideas about sex and sexuality and, once inscribed into nature/nonhuman animals, they apply these values and ideas back to people. The news referred to values, concepts, and features derived from human animal behaviors and human animal society; for example, they quoted the concepts of homosexuality, gays, faithfulness, true love, female persuasion, coercive harassment, sexual self-determination, and happiness. And they superimposed gender stereotypes, like feminized gay men and hypersexual Swedish blonds, onto nonhuman animals. Doing so, the newspapers commingled human animals with nonhuman animal sexualities; in other words, they wrote about nonhuman animals and simultaneously meant human animals. Hence, the newspapers demonstrated and reinforced thinking with nonhuman animals.

The protest of the gay and lesbian communities against the experimental coupling of the gay penguins with female penguins was also telling in this respect. The activists fought for the freedom of all homosexual individuals, belonging to both human and nonhuman animal groups. From the perspective of the gay and lesbian communities, the conversion of the gay penguins into straight ones was equivalent to attempting to force heterosexuality on gay men. The same thinking went for the zoo's director. Her reassurance that they were not trying to heterosexualize the gay penguins indicated that she regarded the charge of heterosexualizing as negative for the zoo's public relations.

She understood that trying to make gay penguins copulate with female penguins might be regarded as fighting homosexuality in general, which seemed to result in bad press for the zoo. Due to the hot news coverage and the subsequent activism of gay and lesbian communities at national and international levels, the zoo could not do whatever it wanted to do with its homosexual nonhuman animals. The director's explanations that the zoo wanted only to rescue the Humboldt penguins from extinction and explore the reasons for and strength of their homosexuality confirmed this impression of public pressure. The director sought to spread the message that the zoo had only good intentions—scientifically and morally.[7]

Another way the news coverage commingled human and nonhuman animal sexuality by thinking with nonhuman animals was to superimpose human animal motivations onto photos of the penguins. For instance, pictures highlighted the idea of happy and cute gay penguin couples leaving females sad and lonely. Two penguins are sitting in front of a cave with news captions such as "Happy without women: The two male penguins 'Vielpunkt' and 'Z' are a couple," "They obviously do not need a woman to be lucky," and "Like to stay with their own: Penguin-men in Bremerhaven's zoo" (Deutsche Presse Argentur, *Stern* 2006; Associated Press, *Stern* 2005; Kiehne 2005). Other pictures showed a lone penguin (presumably female and lonely) apart from two (presumably male) penguins, "Penguin-lady and fellows in Bremerhaven's zoo," or one penguin in front of a cave, looking into it as two penguins lying inside the cave look out, highlighting the female penguin as sad, "Swedish female . . . go on cuddling" (Associated Press, *Sueddeutsche* 2005; Bethge 2005, 187). Those pictures gave both positive (gays are happy) and negative (females left lonely) connotations to the gay penguins' behaviors.[8] An additional stereotype, with mixed messages, imposed onto the gay males was that they were really trying to be females; while described as cute and motherly, the inference was that they were nonetheless stupid since they had been trying to hatch stones and hence were desperately trying to act like females but, of course, failed to be females and to reproduce.

In summary, all groups involved in the public discussion not only conceptualized but also negotiated within heteronormative values and ideas about sexuality through thinking with nonhuman animals. The zoo's experiments with gay penguins, the newspapers' depictions of them, and the reactions of the gay and lesbian communities reflected controversies about homosexuality in human animals and were parts of its discursive negotiation.

Love, Loss, and Family among New York Gay Penguins

In the United States, different newspaper articles featured stories about gay penguin couples over a period of several years. They discussed, for example, that Christians and conservatives liked the documentary *March of the Penguins* for its apparent but subtle Christian imagery and traditional, pro-family messages. The media also covered public debate about the gay penguins in Germany's Bremerhaven Zoo and the objections to the American children's book *And Tango Makes Three* (Richardson and Parnell 2005) (see below).[9] In particular, the stories talked about two specific penguin couples in the United States, Wendell and Cass, two male penguins at the New York Aquarium in Coney Island, Brooklyn (Cardoze 2004), and the most popular penguin couple reported in the media, Silo and Roy in the Central Park Zoo in New York City. The latter two males had spent most of their lives together. After showing courtship behavior and trying to incubate a stone, they got a fertilized egg from their keeper, successfully hatched it, and raised their chick called Tango (Bengiveno 2004; Cardoze 2004; Kuntzman 2005). In September 2005, several articles focused on the fact that Roy and Silo had broken up. After they had lived as a couple for six years, other male penguins aggressively displaced them from their nest, and Silo then bonded with a female penguin.

As in the German news coverage of the Bremerhaven penguins, all reporters used language that mingled human and nonhuman animal lives. Writing about penguins, they mobilized expressions like "soap opera world of seduction and intrigue . . . [and] tales of love, lust and betrayal" (Cardoze 2004); "great lovers . . . really wanted a kid . . . an old married penguin couple" (Kuntzman 2005); "One of America's A-list gay couples has broken up . . . set up housekeeping together. . . . They even adopted a child together. . . . [H]e was wishing for a California girl" (Throckmorton 2005); "Silo's eye began to wander . . . [and] he forsook his partner of six years" (Miller 2005a, 2005b). Pictures highlighted themes of love and betrayal. One article stressed that a female broke up the relationship of Roy and Silo, and the picture showed Silo nuzzling with his new female partner, Scrappy (Miller 2005a). Both image and language evoked betrayal, the two betrayers being caught in the act, but also implied that heterosexuality was stronger than homosexuality.

Cardoze's article told a love story about the two male penguins Wendell and Cass as "one of the best couples at the aquarium," and a picture showed them nuzzling together, underscoring the notion of love. The articles further mingled understandings of penguins' and people's lives by invoking stereotypes of gay men: One penguin is "the more dominating of the two" and "a fierce fighter" that kept other penguins away from their nest, and the other is "afraid of his own shadow" (Cardoze 2004). In Nicole Bengiveno's news article, her title included this: "Love that dare not squeak its name," a play on the phrase which was coined by Lord Alfred Douglas in his poem "Two Loves" (1896), at a time when homosexuality was a criminal offense in England (Bengiveno 2004). Other media articles tended to draw explicit analogies between penguins and people: "Like women in a nunnery or men on a big Navy ship, there's homosexuality in those cases" (*New York Times* 2005). Gersh Kuntzman (2005) alternated throughout his whole article between statements about people and statements about penguins or other nonhuman animals. "Equivalent of going to City Hall in San Francisco," he naturalized same-sex marriages: "If homosexual marriages exist in the animal kingdom, they must be normal" (Kuntzman 2005). He also made points about love and commitment by contrasting the characters in the sitcom *Sex and the City* with the gay penguins: "a First Couple of Monogamy that would show the world that love and fidelity could still conquer all" (Kuntzman 2005). The entire article was a clear example of thinking with nonhuman animals since the author expressed his point of view about (human animal) gay marriages by telling a story about a male penguin couple in their penguin community.

Differences in Media Coverage in Germany Compared to the United States

In contrast to the German news coverage, the American articles raised self-critical and theoretical questions, such as whether it is correct to use the penguins' behavior as a political or moral commodity. Whether overtly rejecting (Bengiveno 2004; Miller 2005a, 2005b; *New York Times* 2005; Throckmorton 2005) or endorsing (Kuntzman 2005) the strategy of drawing upon nonhuman animals to promote certain positions about gender and family politics, all journalists nonetheless thought with nonhuman animals and used the penguin stories for their own purposes. For example, known antigay journalist Warren Throckmorton (2005) reminded readers of a traditional family structure (father, mother, and child) by including two pictures, both showing two adults and one infant—albeit penguin! The text of the first picture specified Roy and Silo looking at and trumpeting to each other with pup Tango between them. The second picture showed the cover drawing of the illustrated children's book *And Tango Makes Three* (Richardson and Parnell 2005), which also showed the three penguins in the same composition as in the first picture. Ironically, these pictures showed non-heteronormative families (two fathers and one child, created by assisted procreation). Throckmorton accused the gay community of taking that penguin couple as a moral role model, while claiming that he himself did not derive lessons about human animal traits from nonhuman animal behavior. His analysis attempted to discredit gay people on that basis, even though he did the same thing himself. Pro-gay Kuntzman (2005) deployed the penguin stories to dispel myths about gay marriages. He argued that Silo and Roy's same-sex partnership proved, first, that gay marriages did not destroy heterosexual marriages and communities and, secondly, that gay marriages were not just about sex since the sexual activity between the male penguins stopped before completion. Bengiveno (2004) argued that sex was not necessarily about procreation and accompanied the article with a picture showing two male penguins looking at each other, almost touching their beaks, presumably lovingly. The picture text highlighted the information that 450 nonhuman animal species were found to exhibit homosexual behavior.[10]

In these U.S. pieces, like the German ones, the penguins' stories served to negotiate legitimate sexualities and family structures in society through language, pictures, and argumentation. Also, like the German coverage, sex bias against females was striking. The articles mainly talked about male penguin couples but did not tell detailed stories about individual female couples, only mentioning that there were also lesbian penguins—including Tango (Miller 2005a)![11] This case study illustrates that sex bias remains a problem within discussions of homosexuality rights.

The public debates about gay penguins in Germany and the United States showed similarities: both intermingled the sexualities of penguins and people by choosing words that expressed thinking with nonhuman animals. In addition, all participants of the debates focused on family, gender, and sexuality politics—albeit in different ways. Thus, very different constituencies of the U.S. and German public (among them, conservatives, Christians, gay and lesbian communities, and pro-gay supporters) negotiated their values about legitimate sexualities and proper family structures with the help of penguins. One striking difference between the German and the U.S. debates was that the U.S. journalists not only used the penguin stories to express their family, gender, and sexuality politics (and to sell newspapers) but also problematized this strategy (which served different critiques about sexuality politics too), whereas the German journalists did not question the strategy itself (see also note 9).

Tango, a Penguin Pup with Two Fathers

The three penguins Silo, Roy, and Tango in the Central Park Zoo in New York City inspired two groups of authors to write children's books about their lives together. The plot of this true story went as follows: The two male penguins tried to breed stones, got an egg from a keeper, and hatched a female that was named Tango.[12] Based on this story, Americans Justin Richardson and Peter Parnell wrote (and Henry Cole illustrated) a children's book, *And Tango Makes Three* (2005).[13] The next year, Edith Schreiber-Wicke and Carola Holland published *Two Daddies for Tango* (2006) in Germany.[14]

Just Like All the Other Families in the Big City

The illustrated book *And Tango Makes Three* gave a short description of well-known Central Park in New York City and its zoo, including all its various animal families. Three human animal families shown entering the zoo included a single woman with a stroller; a man, woman, and child, all holding hands; and two women with two children. A fourth grouping was in the zoo already: an older man (although gender is ambiguous) with a perhaps teenage youth and a younger child; racial-ethnic diversity was suggested across and within the groups. The book then described what "boy" and "girl" chinstrap penguins generally did in order to become a couple. The story continued that, in contrast to the other penguins, the two males, Roy and Silo, did not show interest in the female penguins but did everything together that male and female penguin couples did: swim, sing, walk, and bow to and with each other. They even made a nest of small stones with an empty spot for an egg. Basing their story on the keeper's, whose name and likeness were included in the book, the authors wrote that Silo and Roy looked at what the other couples were doing and imitated them. Their keeper noticed their behavior and thought them to "be in love." Although they patiently tried to hatch a stone, they could not procreate until their keeper gave them, from a reproducing couple, an egg that, from past experience, the keeper knew would not survive. Finally, little Tango hatched out and became "the very first penguin in the zoo to have two daddies" (Richardson and Parnell 2005).[15] As in the beginning of the book, the authors highlighted the aspect of family building by stating at the end, "Just like all the other penguin families . . . and like . . . all the families in the big city."[16]

The book generated significant controversy. Positive reactions applauded the book for teaching tolerance and acceptance of alternative families, highlighted it as a true story (and thus natural), and suggested that a homosexuality that is found in nature/nonhuman animals can protect the rights of gay and lesbian humans (for example, *San Francisco Chronicle* 2005). In contrast, parents in some U.S. states, like Georgia, Tennessee, Iowa, Indiana, Illinois, Missouri, and Wisconsin, were concerned about their children reading stories about gay penguins. They did not want their children learning about homosexuality in nonhuman animals and regarded it as inappropriate for elementary school children. Therefore, they asked schools and public libraries to remove or restrict access to the book (*Boston Globe* 2006). In 2006 (and then for the next two years), the book topped the American Library Association's most challenged books list for a total of 546 formal complaints filed with a library or school in 2006, requesting that materials be removed because of content or inappropriateness (American Library Association 2007).[17]

The book reflected the naturalization of homosexuality and family structures by positioning it in nonhuman animals. In addition, parents' concern that homosexuality was imposed on children by these penguin stories underscored the strength of thinking with nonhuman animals and the belief that children's stories about nonhuman animals mediate social conditions and cultural values.

A Different but Picture-Book Family

The German children's book about the same family, *Two Daddies for Tango*, began by explaining that its story differed from many other penguin stories. It told a story of two male penguins that had always been friends and lived as the "happiest penguins far and wide" (Schreiber-Wicke and Holland 2006). When they were grown up, their two keepers separated them since they wanted them to get interested in females to reproduce. The keepers discussed the question of whether it was good to keep gay penguins and to assist with their family building, explaining, "Two male penguins are not able to have penguin children. You cannot help it. The penguin girls lay eggs and without a penguin egg, no penguin child." Since the separated male penguins refused to eat, one keeper brought them back together against the will of the second keeper and gave them a neglected fertilized egg. All other heterosexual penguins already had hatched small penguin pups and lived as happy families, but the gay penguins needed a lot of time and patience incubating their egg, so the second keeper believed, "It will not happen. Nature did not intend this. Two fathers! Where does something like this happen!" Finally, little Tango hatched, and although her hugging fathers looked strange to her, she danced happily with them. From that time on, these three were "a family like every other. Well, not really like every other family but definitely a happy family. And a picture-book family."

In Internet reviews, the German book generated a good deal of interest and positive reactions but was also controversial. The reviews lauded the book for telling a true story and for referring to nature. Even the Evangelical Institute for Church and Society referred to the story as unnatural and deviant, but added it was a good example of teaching tolerance and uncommon living arrangements to children (Kircheundgesellschaft 2006). In contrast, the conservative Austrian association Soziale Verantwortung (Societal Responsibility) criticized the book for offering the possibility of various sexualities to four-to-five-year-old children; the association also equated homosexuality with the disease of obesity, which they asserted should be fought against rather than supported (Gesellschaftsverantwortung 2006).

The basic message of this book was that, although nature did not usually intend homosexuality, it might happen. Gay penguins were weird and deviant, but they were able to build happy partnerships and—with a little help from some friends—even families. Framed in a heteronormative structure, the children's book described homosexuality as unnatural and deviant, while teaching tolerance and support of it within the ideal of a core nuclear family.

It's All in the Family

These children's books were another significant site for commingling nonhuman and human animal sexualities. Both the parents' concerns and negative reactions, as well as the positive statements of other parents about teaching tolerance, demonstrated it clearly. With the two books' somewhat different messages, the penguins served to teach at least two things to children. First, the books taught children to think with nonhuman animals. Second, they taught which sexualities were legitimate; that is, they instructed children to be tolerant and to accept homosexuality or, alternatively, to regard homosexuality as unnatural but perhaps tolerable.

These children's books reveal that being able to build a family is considered crucial for the acceptance of gays and lesbians and their integration into society. Gay nonhuman animals may be unnatural, but they could build a loving family, which seemed to be the most important basis for accepting nonheterosexual social arrangements. And for building a family, the issue of whether same-sex animals, human or nonhuman, can "reproduce" is significant since human and nonhuman animals can and do "adopt" and rear offspring that are not "theirs" biologically, as the story of Silo, Roy, and Tango illustrates. This issue is at the heart of the meaning of "family" and "parenting" for penguins and human animals alike. According to the critical concept of homonormativity, such a description of a nonheterosexual family was not a sign of liberation or tolerance, but of reactionary "politics that does not contest dominant heteronormative assumptions and institutions but upholds and sustains" (Duggan 2002, 179) normative and family-oriented formations within contemporary economic and political systems.

Thinking (and Making Feminist Change) with Penguins

What have we learned from the public discussion of gay penguins in the news and in children's books? The news coverage of actual gay penguins demonstrated that people interchanged human and nonhuman animal lives by means of language, pictures, and reasoning. Nonhuman animals serve as role models for human animals and help to mediate and negotiate social and cultural values as well as legitimate family structures and sexualities. With children's books, children from an early age already learn to think with nonhuman animals and to practice relations based on beliefs about sexualities and family structures embedded in books. They learn that reproduction and family building are crucial for the integration of homosexual animals into society. We can conclude that to promote tolerance and acceptance, it may be necessary to include in children's books positive stories about non-heteronormative nonhuman animals and social arrangements.[18]

Thus, people are accustomed to thinking with nonhuman animals all the time and, not surprisingly, did it in this instance to assert differing positions on gender, sexuality, and family. Our analysis indicates that homosexuality is still a controversial issue in public opinion in Germany and the United States. Thinking with nonhuman animals, some people regard homosexuality as something to be tolerated, if not supported. Others are afraid that forcing heterosexualization on nonhuman animals might appear to be against homosexual human animals and, as a result, be bad for public relations. In cities like Basel, Zürich, London, and Amsterdam, the visibility and activism of gay, lesbian, and transgendered people were able to initiate changes in the representation of knowledge about sex and sexuality among nonhuman animals. Therefore, such activism was important for societal and scientific developments (cf. Epstein 1996).

Furthermore, an important strategy for acceptance of gays (and lesbians) seems to involve locating homosexuality in the context of reproduction and family building. The abilities to procreate and to build a family appear to be core normalizing concepts that give societal value to variant sexualities in human and nonhuman animals, with non-heteronormative families being somehow, in some way, "a family like every other" (Schreiber-Wicke and Holland 2006). In heteronormative societies, this homonormative strategy is the ticket to acceptability. In our view, this strategy is problematic for several reasons, one being its impact on non-heteronormative families and social relations. The modified but still restrictive definition of "family" may make adoption and foster care by same-sex couples acceptable, but it nonetheless could result in negative implications for childless couples, single individuals, sisters living together, and other social relations that do not fit the traditional heteronormative nuclear family.

Basic arguments for the rights of all individuals regardless of sex or sexuality (or other marks of difference) should not require that heterosexuality, gayness, and queerness be phenomena found in (nonhuman) "nature"; but thinking with nonhuman animals implies a naturalization, which is a valid concern for the practical purposes of social justice work that draws on biological information. The approach of thinking with nonhuman animals and the reference to laboratory animals as role models for human animals has functioned as a tool for social change (or the opposite), reflecting a dynamic between the production of scientific knowledge and societal values (predominant or alternative to predominant). In other words, "nature" is, to a certain degree, flexible; and various cultural values and societal relations can be inscribed into "nature" or nonhuman animals and then, in turn, can legitimize those values and relations. As we saw, gay penguins were used by both antigay and pro-gay proponents. Naturalization has always been a highly problematic strategy, and feminists and others have warned against it. Crucial are the societal and cultural values and social and political relations people want to and choose to live with. They do not need to be negotiated by thinking with nonhuman animals, but since negotiating with nonhuman animals is so powerful, we see advantages in not ceding this strategy to heteronormative politics alone. We want to point out that, although it is problematic, comparisons between human and nonhuman animals can serve queer politics as well, as long as the naturalization processes involved are made clear.

Using the recent spate of public interest in gay penguins in zoos, we illustrated how the approach of thinking with nonhuman animals functioned as a critical negotiation tool in public understanding of human animals in relation to nonhuman animals and vice versa. We argued that thinking with nonhuman animals also helps to explain how stories about nonhuman animals' sexuality function in society, as well as to account for the high degree of interest in nonhuman animal behaviors, particularly surrounding issues of sexuality, family, and social justice.

Sex and sexuality among nonhuman animals provide stories that sell well and, therefore, are typically favored in the media. Public media and institutions use these issues to gain more readers and visitors. In our examples, when public media and educational institutions used cute nonhuman animals, they brought attention to the pros and cons of social justice concerns and important societal values and influenced opinion building (in both directions). Storytelling about nonhuman animal sexual behavior in public media and institutions was used for heteronormative politics, but it also served as a tool to argue for the normality of gender and sexuality variations in human animals and, thus, for queer politics. But using nonhuman animals in order to legitimate human animal affairs and values should always be recognized as making cultural constructions, human animal-inflected stories from science. The understanding of nonhuman animals is culturally shaped, and descriptions of them should not be misunderstood as "truth" or "given in nature." Thus, our examination of gay penguins in zoos illustrated the close intertwining and even co-construction of popular science and societal norms, raising questions about just how objective popular (and even perhaps formal) science can be on topics close to (human animal) home.[19] As *And Tango Makes Three* coauthor Justin Richardson stated, "We wrote the book to help parents teach children about same-sex parent families. It's no more an argument in favor of human gay relationships than it is a call for children to swallow their fish whole or sleep on rocks" (quoted in Miller 2005a).

Whether we extract moral conclusions from penguins eating fish whole, sleeping on rocks, or having same-sex relationships depends on our values, which in turn shape our understanding of science.

ACKNOWLEDGMENTS

We wish to acknowledge University at Albany librarian Deborah LaFond and Oldenburg student assistant Jennifer Glandorf for their invaluable research assistance.

NOTES

1. By "same-sex sexual behavior," we refer to physical sexual intimacy along with social, emotional bonding.
2. See also Franklin, who explored animals as good to "think through or resolve social tensions, conflicts and contradictions . . . (and) good to think about what it is to be properly human" (1999, 10) in modernity and postmodernity. One of the authors (Spanier) believes, however, that Daston and Mitman (2005) neglected the important influence of evolutionary psychology.

3. These programs come from particular zoos and focus on celebrity animals as a way of building on the growing popularity of cable channels devoted to nature, like Animal Planet in the United States, to generate sorely needed publicity for zoos' survival. Several German cable channels also broadcast zoo-soaps throughout the day and evening.
4. *March of the Penguins* (Luc Jacquet) attracted more than 1.25 million viewers and is the most successful animal documentary film in German theaters. The French documentary on the life cycle of the Emperor penguins in the Antarctica outnumbered Wim Wenders's documentary *Buena Vista Social Club* and Michael Moore's documentary hits *Bowling for Columbine* and *Fahrenheit 9/11*. *Happy Feet* brought in 42.3 million dollars and was more successful than the James Bond *Casino Royal* (Deutsche Presse Argentur, *Hamburger Abendblatt* 2006).
5. See, for example, Gorgel 2008, who pointed out that *March of the Penguins* promoted traditional family politics, and Halberstam 2008, who contrasted the contradictory ideology of this story about love and hetero-reproductive family units with a queer reading of the penguin's life in a nonhuman non-monogamous world of affiliation.
6. In 2007, the zoo repeated the experiment once again, but the media did not report it. Kück, the zoo director, said that the male penguins coupled in the same pairs, although the females exhibited some interest in the males. Presumably, it was not interesting enough to the media to tell the same story for a third time (e-mail correspondence in June 2007 by Ebeling).
7. A visit to the Bremerhaven Zoo in 2008 found the zoo dealing openly with the gay penguins. The zoo added a statue of two male gay penguins and information in the front of the penguins' enclosure about two "lesbian" penguins. Moreover, the souvenir shop sold the German children's book *Two Daddies for Tango* (Schreiber-Wicke and Holland 2006). In June 2009 the zoo again got public attention by announcing that two male gay penguins successfully hatched an abandoned egg.
8. Ironically, since humans cannot distinguish male and female penguins by sex (no obvious sexual dimorphism), journalists can ascribe whatever they like to the pictures of penguins; and nobody, neither the readers nor scientists nor the journalists themselves, can assess the veracity of the descriptions superimposed onto photos of a bunch of penguins.
9. The U.S. information about the experiment to "heterosexualize" the gay penguins in Bremerhaven Zoo differs from the German news. The former reported that the zoo stopped the experiments because of protests by gay rights groups, while German newspapers said that the experiments failed, since the gay penguins were not interested in the females. So, while the German media highlighted the penguins' behaviors and described the zoo as a scientific institution that did not indulge the gay activists' protests, the U.S. media gave credit to gay activists for ending the experiments.
10. Bagemihl (1999) included only well-documented scientific observations of mammals and birds. Driscoll (2008) cited "as many as 1,500 species of wild and captive animals," which include all forms of animals, such as insects.
11. It is striking how little attention the press gave to the fact that female Tango grew up to bond with other females! Perhaps the lack of public notice about Tango's homosexuality can be ascribed to prevalent beliefs that gayness is a fixed biological trait, and Tango could not have genetically inherited her "gayness" from the gay fathers who reared her.
12. The Allwetterzoo Münster (Germany) described the same procedure for their gay penguins as typical treatment (Allwetterzoo Münster n.d.).

13. The illustrator, Henry Cole, had worked with the very out, gay activist, writer, and performer Harvey Fierstein to produce *The Sissy Duckling*, an earlier (2002) Simon and Schuster Book for Young Readers and a much less subtle story of a very gay (although not sexual) duckling who finally earns the respect of his highly critical father and peers and is accepted back into his family and community.
14. The book is written in German and has not been formally translated into English; translation here by Ebeling.
15. Neither children's book discussed here has page numbers.
16. We must mention that the book's drawings and text are simply adorable, sweet, and loving.
17. In Great Britain, some parents and churches also protested against this book (Sator 2007).
18. So far, only a few children's books are easily available about queer animals. *And Tango Makes Three* is noteworthy for its basis in a "true," well-documented observation of actual chinstrap penguins. *The Sissy Duckling* (Fierstein 2002) is a heartwarming story about a family and community coming to accept a non-traditional male duck, but it does not claim to be about what ducks actually do.
19. The issue of politics in relation to scientific objectivity is an important one but is beyond the focus of this chapter.

REFERENCES

Allwetterzoo Münster. n.d. http://www.allwetterzoo.de (accessed September 26, 2008).

American Library Association. 2007. www.ala.org (accessed July 30, 2009).

Associated Press, *Stern*. 2005. Schwule Pinguine "Zwangsweise Belästigung durch weibliche Verführungskünste." February 11. stern.de. www.stern.de/wissenschaft/natur/:Schwule-Pinguine-Zwangsweise-Bel%E4stigung-Verf%FChrungsk%FCnste/536441.html (accessed September 26, 2008).

——, *Sueddeutsche*. 2005. Bekehrungsversuch schlägt fehlt. February 9. sueddeutsche.de. www.sueddeutsche.de/panorama/artikel/507/47460 (accessed September 26, 2008).

Bagemihl, B. 1999. *Biological exuberance: Animal homosexuality and natural diversity*. New York: St. Martin's Press.

Bengiveno, N. 2004. Love that dare not squeak its name. February 7. *New York Times*. www.sensualism.com/gay/index.html (accessed September 26, 2008).

Berger, J. 1980. *About looking*. London: Writers and Readers.

Bethge, P. 2005. Frackträger vom anderen Ufer. *Der Spiegel* 7:187–188.

Beyerstein, L. 2006. Gay animal expo draws crowds to Oslo zoo. October 19. alternet.org. www.alternet.org/blogs/peek/43228 (accessed August 28, 2008).

Boston Globe. 2006. Schools chief bans book on penguins: Tale describes males raising egg. December 20. www.boston.com/news/nation/articles/2006/12/20/schools_chief_bans_book_on_penguins (accessed September 27, 2008).

Breadsworth, A., and A. Bryman. 2001. The wild animals in late modernity: The case of the Disneyization of zoos. *Tourist Studies* 1 (1): 83–104.

Cardoze, C. 2004. They're in love. They're gay. They're penguins . . . And they're not alone. July 2. Columbia News Service. http://jscms.jrn.columbia.edu/ (accessed September 26, 2008).

Daston, L., and G. Mitman, eds. 2005. *Thinking with animals: New perspectives on anthropomorphism*. New York: Columbia University Press.

Deutsche Presse Argentur, *Ärzte Zeitung*. 2005. Lockrufe vom anderen Ufer. February 12. *Ärzte Zeitung*. www.aerztezeitung.de/panorama/auch_das_noch/?sid=344388 (accessed September 26, 2008).

——, *Hamburger Abendblatt*. 2006. "Pinguin"—Film erzielt Besucherrekord. January 11. abendblatt.de. www.abendblatt.de/daten/2006/01/11/521878.html (accessed September 26, 2008).

——, *Stern*. 2005. Tierisch: Schwule Pinguine bleiben hart. February 9. stern.de. www.stern.de/wissenschaft/natur/:Tierisch-Schwule-Pinguine/536330.html (accessed September 26, 2008).

——, *Stern*. 2006. Schwule Pinguine: Wahre Liebe nur unter Männern. February 8. stern.de. www.stern.de/wissen/natur/schwule-pinguine-wahre-liebe-nur-unter-maennern-555195.html (accessed September 26, 2008).

——, *Sueddeutsche*. 2005. Homosexuelle Pinguine sollen Nachwuchs zeugen. February 4. sueddeutsche.de. www.sueddeutsche.de/panorama/artikel/280/47233/ (accessed September 26, 2008).

——, *Taz*. 2005. Schwedische Verführung. February 5. taz. www.taz.de/index.php?id=archivseiteanddig=2005/02/05/a0363 (accessed September 26, 2008).

Driscoll, E. V. 2008. Bisexual species: Unorthodox sex in the animal kingdom. *Scientific American*. http://www.sciam.com/article.cfm?id=bisexual-species (accessed September 21, 2008).

Duggan, L. 2002. The new homonormativity: The sexual politics of neoliberalism. In *Materializing democracy: Toward a revitalized cultural politics*, edited by R. Castronovo and D. Nelson, 175–194. Durham, NC: Duke University Press.

Ebeling, K. S. 2002. Die Fortpflanzung der Geschlechterverhältnisse. Das metaphorische Feld der Parthenogense in der Evolutionsbiologie. Mössingen-Talheim: talheimer.

Epstein, S. 1996. Impure science: AIDS, activism, and the politics of knowledge. Berkeley: University of California Press.

Fausto-Sterling, A. 2000. *Sexing the body*. New York: Basic Books.

Fierstein, H., with H. Cole. 2002. *The sissy duckling*. New York: Simon and Schuster Books for Young Readers.

Franklin, A. 1999. *Animals and modern cultures. A sociology of human-animal relations in modernity*. London: Sage Publications.

Gesellschaftsverantwortung. 2006. http://gesellschaftsverantwortung.at/thema_ep/material/FAQ_14_Schule+Aufklaerungsbroschueren.htm (accessed July 27, 2007).

Gorgel, C. 2008. Romantische Liebe, Ehe und Familie in der Tierdokumentation "Die Reise der Pinguine" (Romantic love, marriage, and family in the animal documentary film, *March of the Penguins*). MA thesis, Carl von Ossietzky University of Oldenburg, Germany.

Gowaty, P. A. 1981. Sexual terms in sociobiology: Emotionally evocative and paradoxical jargon. *Animal Behavior* 30:630–631.

Halberstam, J. 2008. Animating revolt/revolting animation: Penguin love, doll sex, and the spectacle of the queer nonhuman. In *Queering the Non/Human*, edited by N. Giffney and M. J. Hird, 267–281. Farnham Surrey: Ashgate.

Haraway, D. 1989. *Primate visions*. London: Routledge.

Heume, W. 2005. Schwule Pinguine erregen weltweit die Gemüter. February 12. *Die Welt*. www.welt.de/print-welt/article424638/Schwule_Pinguine_erregen_weltweit_die_Gemueter.html (accessed September 26, 2008).

Kiehne, J. 2005. Schwedinnen sollen schwule Pinguine auf Trab bringen. February 15. heute.de. www.heute.de/ZDFheute/inhalt/13/0,3672,2259117,00.html (accessed September 26, 2008).

Kircheundgesellschaft. 2006. www.kircheundgesellschaft.de/pdf/2006-papa-liste.pdf (accessed July 27, 2007).

Knowles, K. 2006. Gay animal celebration at London Zoo. July 8. http://www.pinknews.co.uk/news/articles/2005–1929.html (accessed August 13, 2008).

Kück, H. 2006. Pressemitteilung Zoo am Meer Bremerhaven. February 8. www.zoo-am-meer-bremerhaven.de/besucher-infos/aktuelles/detailansicht/browse/2/article/aktuelles-vom-zoo-am-meer/?tx.ttnews%5BbackPid%5D017andcHash=76d3c76a95 (accessed June 12, 2007).

Kuntzman, G. 2005. An unusual love story: Penguins accept same-sex commitments. Why do some people have so much trouble with the idea? February 15. *Newsweek*. www.equal-marriage.ca/resource.php?id=168 (accessed September 26, 2008).

Miller, J. 2005a. New love breaks up a 6-year relationship at the zoo. September 24. *New York Times*. www.nytimes.com/2005/09/24/nyregion/24penguins.html (accessed September 26, 2008).

———. 2005b. Male penguins' breakup at zoo makes a big splash. September 29. *New York Times*. www.detnews.com/2005/nation/0509/29/A12–331450.htm (accessed September 26, 2008).

Naturhistorisk Museum. n.d. Against nature? http://www.nhm.uio.no/against nature/index.html (accessed September 29, 2008).

New York Times. 2005. Editorial. Penguin family values. September 18. www.nytimes.com/2005/09/18/opinion/18sun2.html (accessed September 26, 2008).

Richardson, J., and P. Parnell with H. Cole. 2005. *And Tango makes three*. New York: Simon and Schuster Books for Young Readers.

Roughgarden, J. 2004. *Evolution's rainbow: Diversity, gender, and sexuality in nature and people*. Berkeley: University of California Press.

San Francisco Chronicle. 2005. September 25. http://www.sfgate.com/cgi-bin/article.cgi?file=/c/a/2004/02/07/ MNG3N4RAV41.DTL (accessed July 26, 2007).

Sator, C. 2007. Schwulen-Märchen für britische Schüler. March 13. netzeitung.de. www.netzeitung.de/vermischtes/581091.html (accessed September 26, 2008).

Schiebinger, L. 1993. *Nature's body: Gender in the making of modern science*. Boston: Beacon Press.

Schreiber-Wicke, E., and C. Holland. 2006. *Zwei Papas für Tango* (Two daddies for Tango). Augsburg, Germany: Thienemann.

Spanier, B. 1995. *Im/partial science: Gender ideology in molecular biology*. Bloomington: Indiana University Press.

———. 2000. What made Ellen (and Anne) gay? Feminist critique of popular and scientific beliefs. In *Wild science: Feminist readings of science, medicine, and the media*, edited by J. Marchessault and K. Sawchuck, 80–101. New York: Routledge.

Terry, J. 2000: "Unnatural acts" in nature: The scientific fascination with queer animals. *GLQ: A Journal of Lesbian and Gay Studies* 6 (2): 151–193.

Throckmorton, W. 2005. Silo rains on the penguin pride parade. September 19. Catholic Education Resource Center. www.catholiceducation.org/articles/homosexuality/ho0101.html (accessed September 26, 2008).

Zuk, M. 2002. *Sexual selections: What we can learn and can't learn about sex from animals*. Berkeley: University of California Press.

PART THREE

Categorizing Bodies

8

Intersex Treatment and the Promise of Trauma

IAIN MORLAND

Whenever journalists contact me to say that they are making a documentary about intersex, their first question is always, How often are people born with anatomies that are neither clearly male nor clearly female? It is customary, in not only media coverage but also scholarly discussions of intersex, to begin by listing different types of intersex and their prevalence before analyzing the ethics and politics of their medical management, especially critiques of genital surgery. But such an approach gives several false impressions—that bodies can be separated from discourse; that naming intersex and measuring its prevalence happens outside of discourse; that ethical and political critique brings medicine into discourse from which it is otherwise exterior; that maleness and femaleness are unambiguous most of the time; and that knowing something about how often people are born with certain body parts can inform critical discourse on clinical practice. For these reasons, I am not going to tell you prevalence figures and the medical terms for different types of intersex; both are in dispute (Reis 2007; Sax 2002). Instead, I will tell you about particular anatomies by telling you about the ethics and politics of medical and critical discourse, which is where such anatomies typically are located anyway, together with the surgeries performed on them. In other words, let us suspend the assumption that we can know what intersex is, to explore how and why knowledge about intersex is produced. My standpoint in this project is feminist, but this does not mean that my work is political in a way that clinical practice is not; of course, clinical practice is political too, simply because it takes place in a shared world. Therefore, in my view, the correct response to critiques of medicine is not to make clinical practice apolitical—a fantasy of pure technique, exemplified by prenatal intervention—but to make it right, with all the embeddedness in socially specific values that implies. To this end, parts one and two of my chapter will summarize and analyze conventional practice and its critiques, focusing on the social construction of ambiguity and medical construction of gender; parts three and four will examine and reappraise the specific criticism that medicine has caused trauma for patients. I will draw a contrast between the common deficit model of trauma, according to which traumatization is an unintentional failure, and my argument that medicine has been, conversely and startlingly, traumatic by design.

The Social Construction of Ambiguity

Since the mid-1980s, the Western medical management of anatomies known to clinicians as intersex has come under increasingly multifaceted critique. Medical management has been challenged not only as ethically deficient and technically unsuccessful, but also as an affront to progressive gender politics because of its assumptions about femaleness and maleness. The latter criticism strikes to the very definition of intersex, so it is my starting point in this chapter. Contrary to popular and mythological imagination, human intersex anatomies are not functionally hermaphroditic, but they are characterized by what clinicians have called ambiguous genitalia. By this, clinicians mean usually that a person's external genital appearance is confusing to someone who expects female and male genitalia to be visually discrete. So, for example, a larger-than-average clitoris might look a little like a penis to somebody who believes that clitorises, by definition, are small and that penises, in contrast, are large. Conversely, and for the same reason, a small penis without a urethra might seem clitoris-like. The key point here, on which both medical professionals and their critics generally agree, is that the ambiguity of some people's external genitalia at birth is an effect of social expectations; an intersex anatomy is not physically indeterminate, merely unexpected—a "social emergency," as the American Academy of Pediatrics once put it (2000, 138). However, as I shall explain, traditional clinicians and their critics diverge over how best to respond to such ambiguity and how to do so without causing psychological trauma for people born with atypical sex anatomies.

One response to ambiguity is to stop it. To explain this, consider a further sense in which genitals can be ambiguous, in clinical opinion. External genital appearance can meet expectations, yet the internal genitalia customarily associated with such external features can be absent. For instance, a person might have external genitals entirely congruent with social expectations for a female, but internally—and perhaps unknowingly—they may have undescended testes, rather than the expected ovaries. Although such anatomies have conventionally been diagnosed as broadly intersex rather than specifically ambiguous genitalia, it is important to understand how this diagnosis, too, describes ambiguity and thereby shapes the traditional idea of intersex management as the conversion of ambiguity into certainty. An atypical combination of external and internal genitalia means that a person cannot unambiguously be categorized as female or male (Boyle et al. 2005, 574). It might seem common sense to say that although wholly external genital ambiguity is social, there is in contrast something presocial—natural, even—about a person's sex being ambiguous where internal genitalia are concerned. But it is the ability of medical technology to see inside the body that enables ambiguity to emerge. Therefore, I argue that this type of ambiguity is not really different to the external kind: when medical technology detects gonads such as undescended testes, it materializes the body's interior as a submerged but crucially visible surface. Genetic profiling, which can reveal atypical chromosomal configurations such as XXY or XO, is an extension of the same process. So whether through a contrast between external genitalia and gonads or chromosomes, or between different aspects of the external genitalia, ambiguousness is a way in which bodies can be seen, subject to technologies of visibility as well as anatomical expectations. Like beauty or ugliness, ambiguity is simply not a property of bodies. What is very odd about the medical conversion of ambiguity into certainty, I will suggest, is that it *does* entail changing bodies.

Surgery on intersex genitalia treats ambiguity as a bodily property that it attempts to eliminate. This has been standard practice in Western medicine between approximately the late 1950s and mid-2000s. At the time of writing in 2010, genital surgery hasn't stopped, but its advocacy has been tempered by important clinical guidelines created in 2005, which caution against surgery in infancy without demonstrable physiological benefit (Hughes et al. 2006, 556–557). In this respect, one of the most controversial medical protocols has been the surgical reduction of atypically large clitorises. A procedure explicitly without a physiological goal, such surgery intends exclusively to ensure that there is no uncertainty over whether a person has a penis or clitoris. Similarly, in the case of unexpected contrasts between external genitalia and gonads or chromosomes, surgery has been used to permit a person to be categorized unambiguously as female or male. Significantly, this hasn't meant always doing surgery on the body's exterior to create those genitals usually associated with the gonads or chromosomes within. Rather, bodies have been surgically feminized more often than masculinized, on the grounds that it is technically easier to make genitals that meet expectations about female anatomy. Feminization has sometimes entailed deliberately diminishing physiological function—for example, removing potentially fertile but undescended testes instead of trying to make a scrotum and penis to accommodate and accompany them. In other words, surgery to masculinize genitalia according to social expectations has been done less frequently than feminizing surgery, on the basis that the technical difficulty of creating a sizeable penis and scrotum makes masculinizing surgery more likely to generate genital ambiguity than to stop it, for instance through damage to the urethra. One problem with this protocol is that it overlooks the complex anatomy and varied functions of female genitalia; it supposes that femaleness can be produced simply by taking away tissue, in presumed contrast to maleness (Dreger 1998, 256 n. 33). To put this another way, in the surgical quest for certainty, femaleness and maleness are assumed, in different ways, to *not* be ambiguous.

The key feminist insight that gender is something that one does, rather than an attribute that one has, stands in vexed relation to traditional intersex management. For much second- and third-wave feminism, gender in general is characterized by ambiguity; it is ambiguous not only for people born with unusual anatomies. Rather, in the words of Judith Butler, "gender is always a doing" (1999, 33). This means that gender is a continuous improvisation around social scripts for behavior, emotion, and interaction. Consequently, gender's meanings and effects are open to critical negotiation, as well as to conservative reiteration. Understood like this, feminism is a way of doing gender differently and perhaps without precedent, not a descriptive project that aims to tell the truth about gender within the horizon of contemporary society. The latter project would be fatally uncritical because it would aim for certainty about what gender is, instead of critical optimism about how gender might be done, or even undone. But for some clinicians, the role of medicine is quite consciously uncritical, an accommodation of "what *is* rather than what *should be*," as one has commented (Meyer-Bahlburg, personal communication quoted in Kessler 1998, 120). The comment indicates how medical professionalism and conscientiousness are frequently circumscribed by the assumption that gender is a thing about which one can be certain. This is not to imply that doctors are unprofessional or thoughtless in their handling of intersex. On the contrary, I would argue that even caring and technically competent doctors who manage intersex often fail to recognize that gender is an act of constitutive uncertainty. This analysis is important, because it enables a critique of medicine that doesn't accuse individual doctors of negligence or bad intentions. I have come to regard such a depersonalized critique as vital to constructive dialogue about intersex management. Critique can still be polemical; later in this chapter, I will make a very provocative suggestion about the relation between intersex management and trauma.

Before setting out my claim about trauma, it is essential to explain what feminism and medicine surprisingly have in common. On the one hand, in the light of the profound mismatch I have described between feminist and medical views of gender, feminist science studies has led the way in the critical reappraisal of medicine. This began with Anne Fausto-Sterling's discussion in *Myths of Gender* of the dubious cosmetic and behavioral outcomes of surgical clitoral reduction, followed by Suzanne Kessler's analysis in the feminist journal *Signs* of how intersex management exemplifies gender's social construction, rather than its scientific discovery (the uncovering of "what *is*") as doctors may claim (Fausto-Sterling 1985, 133–141; Kessler 1990). Both Fausto-Sterling and Kessler, a biologist and psychologist, respectively, later published books that critiqued medical management at length, drawing on patient testimonies as well as feminist theory to illuminate the ethical and political shortcomings of genital surgery that aims to foreclose ambiguity (Fausto-Sterling 2000; Kessler 1998). But on the other hand, feminist gender theory has converged with the very theory of gender formation that informs surgery for intersex. This is why surgical protocols remained largely unchallenged by feminist studies for over two decades, until the 1980s. The convergence is as follows: just as feminism has analyzed gender as a doing, medicine has also understood gender to be a process. In both accounts, individuals are born gender-neutral, and becoming gendered is a matter of social interaction with others in culture. The defining difference is that for feminism, gendering is ongoing, so it can be neither accelerated nor completed by surgical intervention. For medicine, gender is in process for only a particular period in infancy, until the age of around three to four years (Money et al. 1955, 289–290). According to that view, infant surgery can indeed expedite gender formation. Put differently, the protocol of genital surgery has been justified on the basis that gender, for everyone, develops from an initial period of ambiguity into eventual certainty. This is an account of social construction shared by feminism up to a point, but it diverges from feminism in supposing that gender's construction starts and ends in early childhood and shouldn't be ambiguous beyond that time.

Conservative Constructivism

Readers may be wondering how genital ambiguity and gender ambiguity are related, if at all. Indeed, critics of medicine have posed this question. Whereas clinicians have traditionally understood genital surgery to be beneficial in curbing ambiguity, feminist commentators like Fausto-Sterling and Kessler have retorted that such surgery is a non-sequitur from the insight that ambiguity is social through and through. In their view, a social problem—if it is a problem—requires a social fix, not a surgical one. But I think that traditional medical management makes two assumptions about the relation between genitals and gender, which strikingly enable surgery to be rationalized *as* a social intervention—a way of fostering the social construction of gender according to expectations about genital appearance. First, ambiguity about genitalia is assumed to provoke ambiguity about gender. Second, certainty about genitalia is assumed to foreclose ambiguity about gender. Despite the appearance in these assumptions of a causal link from genitalia to gender, the second assumption does not follow from the first. This is to say that even if ambiguous genitalia do induce gender ambiguity, it is not obvious that gender ambiguity could be prevented by providing genital certainty. The reason is that the process of making genitals unambiguous—surgery—might itself generate gender ambiguity.

This is quite aside from technical difficulties associated with surgical masculinization. Consider again the removal of undescended testes from a person born with the external genitalia expected for a female. To end up with no gonads at all may make a person feel more sexually "ambiguous" than they might have done without surgical intervention in the first place (Holmes 2008, 154). Moreover, it is possible that the highly unusual experience of genital surgery could make anyone uncertain, to some extent, about their gender, irrespective of surgery's functional and esthetic outcomes.

I would argue further that there may be a causal link in the other direction between gender and genitalia: a remedy for genital ambiguity, if one were needed, might be gender certainty. After all, if ambiguity is social, and if gender is a social construction that includes expectations about anatomy, then being certain about one's gender might dissolve anxieties about one's genitalia, entirely without surgery taking place. As feminism has shown, gender can be done in both conservative and progressive ways, so gender certainty might entail being very sure indeed that one is a woman with a penis, or a man without one. In such cases, one's genitalia would not be experienced as ambiguous, and surgery would be irrelevant. This claim is not as implausible as it may seem, because for most adults in the West, the social construction of gender operates largely independently from actual genital anatomy, which is concealed much of the time. As Hale Hawbecker has explained in his contribution to an important anthology of first-person accounts of intersex, *Intersex in the Age of Ethics*, "I get a kick out of it when a male friend says, 'Hale, you have balls.' I have been tempted to laugh and tell him, 'No I don't, actually, but then I have not really missed them much either'" (1999, 113). One does not need to expose one's genitals in order to be gendered by others in daily social interactions. Everyday experience hereby puts in question medical assumptions about the relation between genitalia and gender.

But paradoxically, everyday experience often doesn't put those assumptions in question at all. Even though doctors who advocate infant genital surgery, and parents who consent to its performance on their children, know from daily life that gender works without reference to genital anatomy—that one can "have balls" without having balls—they do not act on that knowledge when making treatment decisions. It is on this point that the medical idea of gender as a time-limited process, formed and finished in early childhood, offers a crucial rationale for decisions that would otherwise appear highly contradictory. Unlike everyday experience among adults, gender *is* attributed by adults to infants on the basis of infant genitalia, and those external genitalia *are* exposed to others more frequently than in adulthood. This is not simply because of an infant's innate dependency on adult caregivers, but also because of cultural conventions about the timing of, and criterion for, the identification of a newborn's gender. When medical professionals and new parents identify a baby as a boy or a girl, they are really making a claim about its external genital appearance: there's nothing else to go on. The peculiarity of that commonplace identification, which conflates genitals and gender, social expectations and anatomy, is thrown into relief when contrasted to the fact that one would not declare other aspects of a person's identity within seconds of their birth—for example, "Congratulations, it's a liberal" or "It's a heterosexual." In those examples, most clinicians and parents would recognize that identity is something that one does over time and in relation to other people. But being a boy or girl seems to be different. Specifically, it has been understood in medical gender theory as something that is *done to oneself by others.*

The theory of gender underpinning infant genital surgery for intersex holds that because genital appearance is the basis for the attribution of gender by adults to infants, and because an infant's dependency on adults means that those genitalia are exposed relatively frequently (in diaper-changing and bathing especially), infant genitals must look unambiguously either female or male in order for a child to be treated by adults as unambiguously gendered. In turn, being treated as unambiguously gendered is theorized as pivotal to gender formation. In this theory, it is not that the child has a gender which preexists its attribution by others; rather, the attribution of gender *is* the mechanism by which gender forms. In other words, this is a theory of the social construction of gender which, unlike feminist accounts, supposes genital appearance to be paramount (Money et al. 1957, 335). Having balls (anatomically) makes people treat you as if you have balls (socially), so your identity becomes that of someone who has balls (psychologically). What is more, in the medical theory, gender formation has finished by the time a person begins to cultivate everyday non-genital markers of gender difference, such as a hairstyle and the ability to accessorize. The gender theory that underpins intersex surgery in infancy can be summarized as follows: humans are born gender-neutral; gender is formed in early childhood; the social attribution of gender is the mechanism of gender formation; and genital appearance is the basis for that gender attribution. As one doctor has stated, "If we agree that the child born with an intersex condition should be assigned either a male or a female sex and raised unambiguously, the parents must be supported in their difficult educational task by a phenotype concordant with the assigned gender" (Nihoul-Fékété 2005, 25). I call this theory *conservative constructivism* because it holds that current social conventions about the relation between genitalia and gender in early childhood are the means by which gender is constructed, and that it is surgery's task to facilitate such construction.

The Deficit Model of Trauma

I am arguing that successful treatment traditionally means turning the social into the individual. If conservative constructivism seems disagreeable from a feminist standpoint committed to progressive gender relations, this is nonetheless no indication that it is ill-intended. As I will now show, the conversion of the social into the individual has been seen as preventative of trauma, and an article in the journal *Pediatric Nursing* is a useful case in point. Authored in the late 1990s by Katherine Rossiter and Shonna Diehl, nurses in pediatric endocrinology and obstetrics and gynecology at major American hospitals, the article notes but dismisses critiques of traditional treatment; the authors even suggest that parents who refuse surgery on their offspring's ambiguous genitalia commit child neglect (1998, 61). Refusing surgery is perceived by Rossiter and Diehl as neglectful because the diagnosis of genital ambiguity is a "psychological crisis," in response to which infant feminizing surgery can avert "undue psychological trauma" (59–60). Whilst acknowledging that genital surgery has risks (which they do not specify), Rossiter and Diehl claim that surgery can enhance a patient's self-esteem and confidence through the reduction of "ambivalence" in parent-child relations in the short term, and thereby the construction of gender certainty for the patient in the long term (60). In the conservative constructivist view exemplified by these authors, it would be highly unlikely for an affected child to develop an unambiguously female or male gender if its genitalia, left unmodified, provoked in parents and caregivers ongoing ambivalence. Rossiter and Diehl are not the only medical professionals to have regarded such ambivalence as "very traumatic" (Nihoul-Fékété 2005, 24). Inventively, Rossiter and Diehl support the argument for surgery by applying the concept of autonomy not to the child, but to its parents. They reason that autonomy is a socially constructed value, which enables surrogate decision-making by parents. By this, they mean that society allows parents autonomy from doctors in choosing medical treatments for minors. However, because such autonomy is socially constructed, Rossiter and Diehl assert that parental decisions should be based "not only on the present best interest of the child but also on what that means for the larger group into which the child will be socialized" (60). Autonomy in their account turns out to be not very autonomous at all: rather, it is conservative of the social conditions under which it is constructed, namely "the larger context of the personal and societal definitions of what it means to be male or female" (60). So turning the social into the individual is a way of conserving the social, in its current form, through the construction of gendered individuals.

Even though the theory of autonomy that supports Rossiter and Diehl's argument is ethically problematic, the subordination of the patient to parental decisions that channel social norms is claimed by them to be psychologically beneficial. Surgery in infancy for intersex has customarily been defended on the grounds that will not be remembered by the person on whom it is performed (Feder 2002, 314). Rossiter and Diehl put a revealing slant on this presumption by arguing that a child would need "cognitive abilities acquired no earlier than age 12" in order to make a reasoned decision about genital surgery, by which time the maturity that enables an "informed ethical decision" would cause "psychological risk" in the form of an aversive reaction by the child to surgery (1998, 60). This argument precludes meaningful decision-making by the child at any age because it assumes that the child would always choose surgery. The authors fail to consider that a child may decide against surgery and might therefore have no cause for aversive behavior. Rossiter and Diehl's account is interesting because it makes clear an opposition between psychology and ethics that is generally implicit in medical literature on intersex: it is as if good practice would mean distinguishing psychological from ethical concerns, then addressing the former to the exclusion of the latter. I think this opposition is faulty. It overlooks the fact that a patient may suffer psychologically precisely because clinicians and parents do not give due consideration to the ethics of treatment. It also suggests, paradoxically, that the best way to address psychology would be to bypass it by making medical interventions very early in a person's life. To the same end, nondisclosure of infant surgeries to patients in later life, and of the diagnoses that surgeries are intended to manage, has been a widespread policy (Creighton 2004, 44). One might suggest, from a conservative constructivist standpoint, that such nondisclosure is continuous with the everyday ways in which people do gender without consciously reflecting on how it is done. From that standpoint, nondisclosure would be just another aspect of how treatment turns the social into the individual. Contrastingly, conscious reflection on intersex would be traumatizing, in Rossiter and Diehl's view, because it would constitute radical divergence from the usual definitions of "what it means to be male or female." But as I shall explain, systematic nondisclosure is a reason why treatment has created trauma, not averted it. The fantasy, exemplified by Rossiter and Diehl's article, that pre-psychological treatment would trump ethical principles such as informed consent and patient autonomy will also be crucial to my argument that treatment is intentionally traumatic.

Treatment has already been called traumatic by some commentators, but not intentionally so. The medical management of intersex has been criticized not merely for its gender politics, but moreover on the grounds that it produces the very psychological trauma it is intended to alleviate or prevent. I find the latter criticism compelling for two reasons. First, it appeals to medicine's own aims. Consider the words of former patient Arlene, whose surgery in infancy included castration and vaginal construction, but whose diagnosis and medical history were not revealed until the age of eighteen. She has reflected that doctors "built me a vagina so I could have so-called normal sex, and that left me too traumatized to ever want that," which she darkly calls treatment's irony (interview quoted in Brown 2008, 145). The argument that treatment may fail on its own terms is difficult to disregard for even traditionally minded doctors; they might claim that gender is a political matter outside medicine's remit (although they would be incorrect in that claim), but they certainly couldn't say the same about the traumatization of patients. The second value of this critique is that it enables treatment to be contested irrespective of whether one believes gender to be natural or cultural; it isn't necessary to take a reactionary stance against progressive and conservative constructivism alike in order to challenge medicine. This is to say that if medical management is traumatic, it is wrong regardless of whether surgery tries mistakenly to override an individual's naturally occurring gender, or whether it successfully constructs an individual's gender according to questionable criteria. In short, one does not need to know whether gender is natural or cultural in order to hold that treatment causes trauma. For the two reasons I've outlined here, I regard trauma as tactical, "a resource that can be used to support a right," as two cultural commentators have written in a different context—which is not to assert that trauma is all rhetoric, but instead to capture how the felt experience of traumatization can be the basis for patient rights advocacy, deployed through the "social intelligence" of patient advocates (Fassin and Rechtman 2009, 11).

One of the largest and most articulate patient advocacy organizations has been the Intersex Society of North America (ISNA), the work of which demonstrates the two qualities of the tactical use of trauma that I have outlined. Active between 1993 and 2008, ISNA became influential in shaping treatment reform by campaigning for patient rights such as informed consent, rather than focusing on the important but less explicitly transformative project of day-to-day pastoral support. In its early campaigning materials, like the grimly humorous "Hermaphrodites with Attitude" newsletter and T-shirts, the society argued that intersex deserves social recognition alongside maleness and femaleness (Preves 2003, 93, 140). This resonated with contemporary scholarship such as the anthropological collection *Third Sex, Third Gender* and Fausto-Sterling's tongue-in-cheek proposal that there are five sexes, although ISNA never directly argued that there ought to be more than two sex or gender categories (Fausto-Sterling 1993; Herdt 1994). Through collaborations with sympathetic academics, and increasingly with clinicians, the society over time shifted its focus from gender, stating significantly in 2003 that intersex "is primarily a problem of stigma and trauma, not gender" (Chase 2003, 240). This shift aligned with emerging research in the medical humanities that, while not about intersex, argued for clinical attention to patients' narratives of their subjective experiences of health and illness (Greenhalgh and Hurwitz 1998); likewise, ISNA's assertion that intersex is not a problem of gender acknowledged that affected individuals—rather than their parents or doctors—are experts on their own genders. In the light of this, the society advocated making a provisional decision about a newborn's gender without doing surgery, then allowing the developing child to articulate its gender and, if necessary, modifying one's decision to fit—rather than modifying the child to fit the decision (Chase 2003, 241). The aim of this recommendation was to bring medical practice into line with the everyday, adult understanding of gender.

Of course, the terms by which any child articulates its gender are culturally organized—the existence of different types and colors of toys, for example—just as the perception of Hawbecker as someone who "has balls" depends on the values associated with masculinity. But ISNA was not eschewing the feminist project of changing the terms of gender, merely suggesting that such a project was extraneous to the provision of what it called "patient-centered care" (Chase 2003, 240). The revelation in the late 1990s of the outcome of a medical test case for gender's social construction substantiated this view. Under the direction of the architect of treatment protocols for intersex, twenty-three-month-old David Reimer was surgically reassigned female in 1967 following injury in a circumcision accident. Despite claims in numerous influential medical publications that he had been thoroughly socialized into being a girl, it later emerged that Reimer had roundly rejected the reassignment and gone on to live as a man (Diamond and Sigmundson 1997). Whereas, for some commentators, this outcome seemed to reveal gender to be natural rather than cultural and, thereby, to disprove feminist accounts of gender as a doing, for ISNA the case exemplified solely the damage caused by secretive childhood surgery and a failure of attention to Reimer's subjective experience (Morland 2007, 89–95). The society's interpretation of the Reimer case was informed by a remarkable paper by ISNA supporter Tamara Alexander, which pointed out that traditional medical management is arguably an "analogue for childhood sexual abuse," most acutely in the perception by infants of treatment as destructive regardless of its good intentions. Alexander suggested that "genital procedures in childhood," because of crossing bodily boundaries, "may have the same affective valence" as abuse (1997). In short, the position of the society and its growing number of clinical allies has been that traditional treatment does not construct gender in the first place, but often inadvertently creates trauma and thus fails by its own standards (Creighton and Liao 2004, 661). The tactical use of trauma, then, inverts the conservative constructivist view that gender can and should be constructed in the service of trauma's prevention.

By inverting each other's positions, traditionalists and their critics share what I call a deficit model of trauma. In this model, trauma is a failure, and it happens principally because of a failure to do something. Just as for traditionalists, not doing surgery is presumed to be traumatic, so too for critics, trauma occurs when an intersex diagnosis or its treatment are undisclosed, or a patient's rights to bodily integrity and informed consent are not honored. Both positions assume that trauma happens unintentionally: one might mean well by not doing surgery, or by nondisclosure, or even by violating informed consent to change an infant's genitals, but trauma will result regardless. And if treatment is subjectively an "analogue" for traumatic abuse, as the evidence gathered assiduously by Alexander indicates, it might appear obvious that this is a measure of treatment's failure, by technical and ethical standards alike. So surgery can certainly be traumatic, as Arlene's comment about vaginal construction testifies, but my point is that in the deficit model, surgery's trauma arises through a failure to recognize that psychological support, not surgery, was necessary to begin with; Arlene is not suggesting that surgery was performed with a view to her traumatization. I am therefore using the phrase "deficit model" to describe a way of thinking about the relation between trauma and intentionality that assumes nobody would act purposely to cause trauma. But in the final part of my chapter, I want to suggest that in the traditional medical management of intersex, traumatization has happened not by omission but by design.

Trauma by Design

In the deficit model that I have set out, trauma can interfere with gender development but is essentially separate from it. Hence, ISNA's claim that intersex is a problem of trauma, not of gender, implies that to reveal treatment as traumatic is to break from the past—in other words, that in the past, the traditional medical approach did assume intersex to be a problem of gender, and this assumption was wrong (Williams 2002, 455). The corollary of the society's claim is that trauma, properly understood, is not a matter of gender. It should be clear that I do not question the tactical value of either implication in campaigning for progressive treatment reform. But I think the role of trauma in conservative constructivism deserves a closer look. What if treatment, specifically in its aim to construct gender, were *traumatic by design?* To ask this question is not to insinuate that clinicians are malevolent. Quite the reverse; I think it is helpful to temporarily suspend judgment about the ethics of psychological trauma in order to examine how traumatization, despite its negative connotations, may be an essential aspect of traditional treatment. It is not a failure by omission. I will argue that what clinicians call the therapeutic construction of gender and what critics of medicine call the traumatization of patients are the same thing. My claim is that treatment protocols, according to medicine's own criteria for success, have constructed gender by traumatizing patients.

The critique of intersex management as traumatic often characterizes trauma imprecisely, as various psychological or emotional aspects of treatment and its effects that are aversive or upsetting. I find this to be problematic. For example, a twenty-two-page article by a family therapist, titled "Intersexuality in the Family: An Unacknowledged Trauma," does not actually define trauma; it indicates somewhat vaguely that nondisclosure "can have an untoward psychological impact" (Lev 2006, 39). Others have identified as traumatic the necessity of repeated surgeries, sometimes simply to repair problems caused by other surgeries, long beyond the eighteen-month period hypothesized as optimal in the original clinical protocols. In these instances, awareness of an ongoing "surgical maelstrom," as one medical team puts it, can be just as upsetting as the belated discovery of nondisclosure (Holmes 2008, 56; Stecker et al. 1981, 539). But in an extensive literature review of scholarship on intersex, the only analysis I have found within a substantive framework for understanding trauma is a four-page discussion in an American Psychological Association (APA) handbook for the culturally literate treatment of posttraumatic stress disorder. Perhaps ironically, the book's brief discussion of intersex is located in a chapter on sex and gender (Brown 2008, 142–145). Both the APA handbook and the family therapy essay make cogent recommendations about psychological support for affected individuals and families, but I want to signal that one risk of using the term "trauma" imprecisely in this context is that it can serve to psychologize political objections to medical treatment, including those expressed by patients (Roen 2008, 60). That is, such objections can appear to be treatment's "untoward psychological impact," rather than the articulation of a reasoned standpoint.

Further, if objections to treatment can seem untoward, then successful treatment may appear to be characterized by the absence of psychological impact. Rossiter and Diehl's article demonstrated this. However, I submit that treatment does seek to have a psychological impact, indeed an utterly overwhelming one, and in this respect its aim is nothing less than traumatization. Individual trauma, according to a classic study by sociologist Kai Erikson, is "a blow to the psyche that breaks through one's defenses so suddenly and with such brutal force that one cannot react to it effectively" (1976, 153). In the light of Erikson's definition, my reasoning is as follows. In traditional treatment protocols, early childhood is theorized to be a time of unprecedented mental "plasticity," uniquely amenable to gender's "imprinting" (Money et al. 1957; Morland 2007, 86–89). Plasticity here connotes resilient adaptability, but it could equally be understood as profound vulnerability (Brown 2008, 117). In other words, infantile receptivity to gendering is intensely ambivalent, a sensitivity to intervention interpretable as both the capacity to absorb impacts without trace and the condition of being shaped irrevocably by those very impacts. Because of such ambivalence, doctors and parents have rationalized infant treatment as at once life-changing (in imprinting gender) and not life-changing at all (in having no impact other than the imprinting of gender). This contradiction can be explained if treatment is analyzed as traumatic by design. An overwhelming blow to the infant psyche, treatment is intentionally sudden and incomprehensible. Surgery epitomizes this. It aims specifically to make an impact for which an individual is unequipped to cope, insofar as the infant has neither the psychological nor interpersonal resources needed to gain critical distance on surgery and thereby to reframe its meaning as objectionably injurious (Brown 2008, 118). Only in infancy can one "make a change of sex with impunity," advises a classic medical paper (Money et al. 1955, 290). Gender is purportedly imprinted by such treatment because the infant, in Erikson's words, "cannot react to it effectively" and, instead, is promptly and permanently changed in a way that forecloses the possibility of conscious reaction and contemplation. Hence, Rossiter and Diehl's recommendation that treatment should predate the development of a patient's "cognitive abilities."

I am arguing that treatment protocols based on conservative constructivism function traumatically when carried out technically correctly and with kind intent. This disturbing incongruity is exemplified by the policy of nondisclosure, which has supposed that surgery makes an impact on gender by not making an impact on memory. In this way, nondisclosure of surgery in infancy generates a key characteristic of trauma: the unrecognized persistence in the present of an incomprehensible past event, which in the words of one cultural theorist, "never becomes part of the ordinary memory system" (Leys 2000, 298). To be gendered by surgery is, by design, to be gendered without knowing that surgery is the cause. To this end, the nondisclosure policy has structured not merely interactions between parents and their children—whereby parents have strenuously obscured formative information—but interactions between doctors and parents too, in which doctors have sought to minimize parental perception of gender ambiguity by withholding the finer details of their child's diagnosis (Feder 2002, 294–295). I identify such nondisclosures as occasions where trauma is actively produced by medical protocols, as opposed to instances of treatment's traumatic failure, which the deficit model would suggest. Consequently, it is important to register that infant genital surgery, as an exemplary traumatic experience, differs from other aspects of treatment in degree rather than kind: parental reassignment of gender on the basis of postsurgical genital appearance is correspondingly abrupt and irrefutable, and equally integral to treatment. Irrespective of whether the child's anatomy is traumatic for the parents, my point is that the parents are a vehicle for the medical traumatization of the child. Nondisclosure by parents generates what Sherri Groveman, a patient support group organizer and Jewish American, calls "a personal holocaust," by analogy with the historical event often considered to be quintessentially traumatic (1999, 27). So I think that former patients like Groveman, who have gone on to criticize medicine, aren't doing so because of untoward psychological sequelae—as some doctors have implied—but, conversely, because they have worked through trauma and thereby achieved sufficient critical distance on surgery in order to reappraise its meaning. As one forward-thinking medical team has noted, being a peer support group member could be regarded as a sign of personal resourcefulness, not of psychological problems (Minto et al. 2003, 162).

In summary, then, I am not saying that treatment is traumatic because it damages a person's innate gender. From a feminist standpoint that regards gender as a social doing, one cannot assume gender to be present in a newborn and susceptible to surgical alteration, diminution, or loss. Instead, I follow the feminist philosopher Candace Vogler in regarding trauma "as getting more than you can handle rather than losing something that you already had" (2004, 41). The peculiarity of traditional treatment on the basis of conservative constructivism is that it tries to construct gender too fast, all at once, traumatically. It is worth noting that this analysis also provides a way to understand adult sexual dysfunction caused by surgery in infancy, such as that reported by Arlene. In my opinion, surgery doesn't impair a natural sexuality that is separate from the social; rather, because sexuality is always social, it is constitutively sensitive to external stimuli and, hence, vulnerable to overwhelming experience (Seidman 2003, xiv). To close, I wish to return to what I described in my introduction as the fantasy of apolitical clinical practice, pure technique, immune from critique. Morgan Holmes, a patient advocate and scholar, has recounted a discussion with surgeons at a medical school in the 1990s. Holmes argued that the failure to disclose the nature or purpose of infant surgery "formed part of the resulting emotional trauma" (2008, 166 n. 21). The surgeons' response to this deficit model argument was that with technical advances, their interventions could take place so early in life, creating so few scars and complications, that disclosure to patients would be superfluous. Indeed, surgery sometimes has been performed within seventy-two hours of birth (Holmes 2008, 131). More recently and in the same spirit, some clinicians have recommended laser soldering in genital surgery on the grounds that it can be done "in almost sutureless fashion and more rapidly than conventional suturing" (Kirsch et al. 2001, 574). If this is the promise of pure technique, I consider it indistinguishable from the promise of pure trauma: leaving no marks, taking no time, it marks an individual for his or her lifetime with the imprint of gender.

ACKNOWLEDGMENTS

I thank Sarah Creighton, Paul Crosthwaite, Ellen Feder, Laura Gregory, Bo Laurent, and Claire Nihoul-Fékété for helpful discussions and information.

REFERENCES

Alexander, T. 1997. The medical management of intersexed children: An analogue for childhood sexual abuse. http://www.isna.org/drupal/node/view/159 (accessed May 16, 2010).

American Academy of Pediatrics. 2000. Evaluation of the newborn with developmental anomalies of the external genitalia. *Pediatrics* 106:138–142.

Boyle, M. E., S. Smith, and L.-M. Liao. 2005. Adult genital surgery for intersex: A solution to what problem? *Journal of Health Psychology* 10:573–584.

Brown, L. S. 2008. *Cultural competence in trauma therapy: Beyond the flashback.* Washington, DC: American Psychological Association.

Butler, J. 1999. *Gender trouble: Feminism and the subversion of identity.* 2nd ed. New York: Routledge.

Chase, C. 2003. What is the agenda of the intersex patient advocacy movement? *Endocrinologist* 13:240–242.

Creighton, S. M. 2004. Long-term outcome of feminization surgery: The London experience. *BJU International* 93 (supplement 3): 44–46.

Creighton, S. M., and L.-M. Liao. 2004. Changing attitudes to sex assignment in intersex. *BJU International* 93:659–664.

Diamond, M., and H. K. Sigmundson. 1997. Sex reassignment at birth: Long-term review and clinical implications. *Archives of Pediatric and Adolescent Medicine* 151:298–304.

Dreger, A. D. 1998. *Hermaphrodites and the medical invention of sex.* Cambridge, MA: Harvard University Press.

Erikson, K. T. 1976. *Everything in its path: Destruction of community in the Buffalo Creek Flood.* New York: Simon and Schuster.

Fassin, D., and R. Rechtman. 2009. *The Empire of trauma: An inquiry into the condition of victimhood.* Translated by R. Gomme. Princeton: Princeton University Press.

Fausto-Sterling, A. 1985. *Myths of gender: Biological theories about women and men.* New York: Basic.

———. 1993. The five sexes: Why male and female are not enough. *Sciences,* March/April, 20–25.

———. 2000. *Sexing the body: Gender politics and the construction of sexuality.* New York: Basic Books.

Feder, E. K. 2002. Doctor's orders: Parents and intersexed children. In *The subject of care: Feminist perspectives on dependency,* edited by E. F. Kittay and E. K. Feder, 294–320. Lanham, MD: Rowman and Littlefield.

Greenhalgh, T., and B. Hurwitz, eds. 1998. *Narrative-based medicine: Dialogue and discourse in clinical practice.* London: BMJ.

Groveman, S. A. 1999. The Hanukkah bush: Ethical implications in the clinical management of intersex. In *Intersex in the age of ethics,* edited by A. D. Dreger, 23–28. Hagerstown, MD: University Publishing Group.

Hawbecker, H. 1999. "Who did this to you?" In *Intersex in the age of ethics,* edited by A. D. Dreger, 110–113. Hagerstown, MD: University Publishing Group.

Herdt, G., ed. 1994. *Third sex, third gender: Beyond sexual dimorphism in culture and history.* New York: Zone.

Holmes, M. 2008. *Intersex: A perilous difference.* Selinsgrove, PA: Susquehanna University Press.

Hughes, I. A., C. Houk, S. F. Ahmed, P. A. Lee, and Lawson Wilkins Pediatric Endocrine Society/European Society for Paediatric Endocrinology Consensus Group. 2006. Consensus statement on management of intersex disorders. *Archives of Disease in Childhood* 91:554–563.

Kessler, S. J. 1990. The medical construction of gender: Case management of intersexed infants. *Signs* 16:3–26.

———. 1998. *Lessons from the intersexed.* New Brunswick, NJ: Rutgers University Press.

Kirsch, A. J., C. S. Cooper, J. Gatti, H. C. Scherz, D. A. Canning, S. A. Zderic, and H. M. Snyder III. 2001. Laser tissue soldering for hypospadias repair: Results of a controlled prospective clinical trial. *Journal of Urology* 165:574–577.

Lev, A. I. 2006. Intersexuality in the family: An unacknowledged trauma. *Journal of Gay and Lesbian Psychotherapy* 10 (2): 27–56.

Leys, R. 2000. *Trauma: A genealogy.* Chicago: University of Chicago Press.

Minto, C. L., L.-M. Liao, G. S. Conway, and S. M. Creighton. 2003. Sexual function in women with complete androgen insensitivity syndrome. *Fertility and Sterility* 80:157–164.

Money, J., J. G. Hampson, and J. L. Hampson. 1955. Hermaphroditism: Recommendations concerning assignment of sex, change of sex, and psychologic management. *Bulletin of the Johns Hopkins Hospital* 97:284–300.

———. 1957. Imprinting and the establishment of gender role. *American Medical Association Archives of Neurology and Psychiatry* 77:333–336.

Morland, I. 2007. Plastic man: Intersex, humanism, and the Reimer case. *Subject Matters* 3 (2)/4 (1): 81–98.

Nihoul-Fékété, C. 2005. Does surgical genitoplasty affect gender identity in the intersex infant? *Hormone Research* 64 (supplement 2): 23–26.

Preves, S. E. 2003. *Intersex and identity: The contested self.* New Brunswick, NJ: Rutgers University Press.

Reis, E. 2007. Divergence or disorder? The politics of naming intersex. *Perspectives in Biology and Medicine* 50:535–543.

Roen, K. 2008. "But we have to *do something*": Surgical "correction" of atypical genitalia. *Body and Society* 14:47–66.

Rossiter, K., and S. Diehl. 1998. Gender reassignment in children: Ethical conflicts in surrogate decision making. *Pediatric Nursing* 24:59–62.

Sax, L. 2002. How common is intersex? A response to Anne Fausto-Sterling. *Journal of Sex Research* 39:174–178.

Seidman, S. 2003. *The Social Construction of Sexuality.* New York: Norton.

Stecker, J. F., C. E. Horton, C. J. Devine, and J. B. McCraw. 1981. Hypospadias cripples. *Urologic Clinics of North America* 8:539–544.

Vogler, C. 2004. Much of madness and more of sin: Compassion, for Ligeia. In *Compassion: The culture and politics of an emotion*, edited by L. Berlant, 29–58. New York: Routledge.

Williams, N. 2002. The imposition of gender: Psychoanalytic encounters with genital atypicality. *Psychoanalytic Psychology* 19:455–474.

9

The Western "Lesbian" Agenda and the Appropriation of Non-Western Transmasculine People

SEL J. HWAHNG

In July 2008, Evelyn Blackwood, one of the two organizers for a conference (the other was Saskia Wieringa) sent out a flyer over e-mail announcing an International Female Masculinities Symposium to be held at the University of Amsterdam in early September 2008. What was noticeable about this flyer was the conflation of "female masculinities" with "gender-ambiguous women." When I pointed out to Blackwood this problematic conflation in that Blackwood herself had published a book chapter (1999) in which "gender-ambiguous" Indonesian transmasculine individuals did not identify as women, but rather as men, Blackwood subsequently changed the flyer so that "female masculinities" became synonymous with "gender ambiguity in female-bodied individuals."

The organizers of this conference failed to explain what their criteria were for "female-bodied individuals" and why "gender ambiguity in female-bodied individuals" was being singled out from the rest of the transmasculine spectrum as a focus for their conference. Since Blackwood and Wieringa focus on non-Western populations that are often low income and/or working class, and specifically on Asian people, is there some inherent fundamental subjectivity among "female-bodied individuals" in non-Western/Asian contexts? If so, what is the evidence for this?

My intention for focusing on Blackwood and Wieringa's research and scholarship as a case study is twofold.[1] First, their research and scholarship exhibit some of the fault lines that are endemic to some of the research and scholarship on anatomically-female and transmasculine individuals, both in the United States and transnationally.[2] Second, Blackwood and Wieringa's tendencies to idealize gender ambiguity in non-Western/Asian "female-bodied individuals" are paragons of Western projections onto non-Western/Asian individuals, in which the actual subjectivities and agencies of these non-Western/Asian individuals become subsumed under a Western gaze that overdetermines and fetishizes non-Western/Asian "bodies." Blackwood and Wieringa also demonstrate that despite their presumably good intentions, even researchers trained in disciplines focusing on cross-cultural inquiry, such as anthropology, may not be able to successfully circumvent Western indoctrination.

As will be discussed in this chapter, Blackwood and Wieringa overwhelmingly rely on gender categories that are based on a Western biological-based definition of bodies. They do not consider if non-Western transmasculine individuals want to be categorized by their "bodies," nor does it appear that non-Western social contexts inform the criteria upon which their bodies are judged and measured. There are four hypotheses in this chapter, which I will attempt to prove:

1. Blackwood and Wieringa operate from within *this* Euro-Western sex/gender system: *anatomy* = *materialized sex (female versus male)* = *gender (women versus men)*. But the non-Western individuals they study operate within *this* sex/gender system: *anatomy (X) social roles* = *materialized sex (often 3 or 4 sexes)* = *gender roles (often 3 or 4 genders)*.
2. Blackwood and Wieringa extend their binary gender oppositionality (women versus men) to also binarize nonheterosexual gender and sexuality (see table 9.1).

TABLE 9.1

Binarization of nonheterosexual gender and sexuality

Women	*vs.*	*Men*
Lesbians		Gay men
Masculine females		Homosexuals
Transgendered females		Queer individuals
		Third-sex individuals

3. Blackwood and Wieringa occupy positions of much more status and socioeconomic privilege (race, class, education, and citizenship) than the non-Western individuals they study, and they impose their Euro-Western sex/gender paradigm onto these non-Western individuals through a series of ideological domination and colonization.
4. Blackwood and Wieringa pursue a series of digressions that appear to complicate their sex, gender, and sexuality categories, yet they repeatedly foreclose any movement toward complexity and reify their reductive Euro-Western paradigm.

Blackwood and Wieringa have co-edited two book volumes, *Female Desires: Same-Sex Relations and Transgender Practices across Cultures* (1999) and *Women's Sexualities and Masculinities in a Globalizing Asia* (with Abha Bhaiya, 2007). What becomes apparent from reading their scholarship is the presentation of often vague and contradictory statements and ideas, in which questionable conclusions are made with either little or no evidence. Furthermore, there appears to be a suspect intention of appropriating supposed evidence drawn from non-Western/Asian individuals and communities in order to reify an agenda that is actually based on a U.S./Western "lesbian" model, in which the categories of "female," "woman," and "lesbian" become naturalized and conflated together. Non-Western/Asian individuals are thus analyzed and filtered through this lens, instead of being systematically studied within their social contexts by employing rigorous methodologies, such as inductive analysis and grounded theory, which would actually observe patterns from the information collected and develop hypotheses, theories, and frameworks *emerging* from this information (Glaser 1992). This would result in "theoretical sensitivity" (Glaser 1978) that is more precisely based on their actual lives.

"Female" Contradictions

In *Female Desires* Blackwood and Wieringa claim that applying Eurocentric categories to non-Western women's sexual practices is problematic (1999, xi) and that "this anthology seeks to defuse the dominance of the West . . . and to expand theories of sexuality beyond the cultural problematic of the West" (2). They also state in *Women's Sexualities and Masculinities* that "making any of the terms [such as lesbian, women, and transgendered] . . . a global category is problematic because it imposes a 'Western' understanding of sexuality, with its reference to a fixed sexual identity, on practices and relationships that may have very different meanings and expectations in other places. . . . Any single solution has its particular consequences, but it is incumbent on researchers to avoid imposing meanings through the unthinking use of Western categories" (Blackwood et al. 2007, 6–7).

In light of such statements, it is surprising that Blackwood and Wieringa succeed in doing exactly that in their scholarship—applying their own Eurocentrically-informed categories to non-Western anatomically-female and transmasculine people's sexual, as well as gender, practices. They also reify and prioritize sexuality and gender as the defining aspects of these non-Western individuals' identities.

Although "female" is in the title of their edited volume, they demonstrate a very contradictory relationship to this term. It appears that they operate from a biologically essentialist definition of "female" yet realize that a direct statement as such would not be considered intellectually rigorous or useful. Blackwood and

Wieringa thus express in the preface to *Female Desires* that "what is common to all these case studies are *female* bodies (this, however, does not mean that the latter necessarily share any predetermined sets of traits, feelings, or experiences)" (1999, ix). At first it appears that they seek to denaturalize and complicate the category of "female." However, toward the beginning of the introduction to *Female Desires* they seemingly contradict themselves by then stating, "We focus on bodies marked as female. . . . Thus we start with material female bodies that are physically marked by genitals and reproductive capabilities and examine the consequences of that marking" (7). This latter statement contradicts the first statement, in that if the category of "female" does not share any *predetermined* sets of traits, feelings, or experiences, it now appears that there indeed are traits that determine femaleness—that of genitals and reproductive capabilities.

This definition of female is highly questionable because it eerily reproduces heteropatriarchal capitalism and takes for granted capitalism's biological paradigm, in which sexual identity is redefined in the idiom of anatomical destiny and biological reproduction is a point of focus and naturalized as a measure of successful production (Morris 1994, 32). Another dubious aspect is that it eliminates many anatomically-female women-identified individuals who have no exchange value within the biological reproductive paradigm (see Fisher, this volume).

The conundrum of incorporating trans/gender-variant individuals within this definition further highlights the questionable aspects of their definition.[3] For many, if not most, transmasculine people who undergo testosterone supplementation, this supplementation curtails their menstrual cycles, and thus they lose their reproductive capabilities. Transmasculine people may also undergo sex reassignment surgery to alter their genitals. It appears that in both cases these anatomically-female trans/gender-variant people would also be excluded from their definition. In considering transfeminine (male-to-female trans/gender-variant) individuals, on the other hand, who undergo sex reassignment genital surgery, which results in female genitals—are they only considered partial females because they still do not have reproductive capabilities? Blackwood and Wieringa do not grapple with any of these questions in their biologically essentialist definition, although they repeatedly invoke the term "transgender" in their scholarship throughout both edited volumes.

Blackwood and Wieringa then undertake an exegesis of an assorted mixture of scholars and theoretical frameworks, some of which critique biologically essentialist definitions of sex. These include Simone De Beauvoir, Linda Nicholson, Margaret Mead, Luce Irigaray, Hélène Cixous, the U.S. "radical feminist" movement of the 1970s and 1980s (including Ann Ferguson, Gerda Lerner, and Christine Delphy), Western feminist social constructionist theories in relation to biblical references of women, Gayle Rubin, Monique Wittig, Judith Butler, Kathleen Barry, Adrienne Rich, the development of the social construction theory of gay/lesbian identity in the United States and Europe, and Michel Foucault as a corrective to biological determinism. They discuss their familiarity as if to demonstrate their expertise with Western feminist theories, with no other actual applicability to the study of non-Western populations. They manifest an ambivalent relationship to most of these theories—not actually incorporating them but not critically refuting them either.

After this exegesis, Blackwood and Wieringa then write: "In such cases female bodies have cultural meaning and significance as *potential reproducers*. . . . Consequently, the female body retains its ideological significance. . . . We argue that the female body is the *source of identity* among all these different categories of sexuality and gender, whether lesbian, transgender, femme, or butch. Female bodies thus constitute the unifying concept for the work in this volume" (1999, 25, emphasis mine).

Again, it is the biological reproductive potential that is emphasized and that lends meaning and significance to the definition of "female." What Blackwood and Wieringa do not grapple with is how they account for their definition in light of the social constructionist and discursive feminist theories that were expounded upon, nor how these theories are relevant to the examination of non-Western individuals.

"Women" and Binary Gender Oppositionality

Parallel to the contradictions and reversals that mask the biologically determinist definition of "female" are contradictions and reversals that mask a binary gender system—that of "women" and "men." In addition, further examination reveals that Blackwood and Wieringa's conceptual framework is also invested in upholding a binary gender oppositionality that is not substantiated by actual evidence from the study of non-Western individuals, but rather appears to be deductively imposed onto non-Western individuals and communities.

For instance, in the introduction, Blackwood and Wieringa state, "We raise the questions about the relevance of theories on women's oppression to studies of female sex practices and ask whether the category 'woman' is inclusive or exclusive of the experiences represented here" (1999, 3). This appears to be a highly critical question to their project that is also supplemented by another question later in the chapter. "The cases presented in this anthology destabilize the idea that lesbians, transgendered females, and women can all be subsumed under one category 'woman.' The proliferation of lesbian and transgender identities suggests that neither gender nor sexuality are unifying markers cross-culturally. . . . If the gender category 'woman' is unstable, is there a reason to distinguish between female and male bodies theoretically?" (25).

They resolve their questions by generalizing that "gender transgression" brings different experiences for those with an "originally female or male body,"

TABLE 9.2
Binarization of non-Western/Asian identities

Female	vs.	Male
Women/*Dees*	vs.	Men
Transmasculine people/*Toms*	vs.	*Bancis/Waria*

proffering an example of Indonesian anatomically-male, female/feminine-identified *bancis* (also known as *warias*) as more "free" than Indonesian trans-masculine people, dismissing the complex identity of *bancis* as third-gender individuals and any potential identifications *bancis* may have as women (Oetomo 2000). They also provide no citations to support their claims. In this example, then, transmasculine people are on the side of "women," while *bancis* are on the side of "men," according to the model shown in table 9.2.

Blackwood and Wieringa then reify the biologically essentialist definition of "female," asserting that the potential biological reproductive capability of the anatomically-female body is what threatens society—which is construed as an abstract, generalized society—and is also the source of identity for all anatomically-female and transmasculine people (1999, 25). All identities, then, are subsumed under the category "female." However, do they make a distinction between the categories of "female" and "women," since they state above that they are actually "destabilizing" the idea that a variety of homosexual and transmasculine identities can be subsumed under one category of "woman"?

Apparently Blackwood and Wieringa do not draw a distinction, nor are they destabilizing this idea, for in another contradictory move in "Globalization," their introductory chapter to *Women's Sexualities and Masculinities*, they state,

> This collection maintains the orientation of . . . *Female Desires* (1999) by including work only on women and transgendered and masculine females. In an era of proliferating "lesbian and gay" studies and studies of queer sexualities that do not attend to the specificities of gendered practices within and across sexualities, we feel it is critical to highlight the particular regulatory practices, state and religious ideologies, and global processes that collaborate with and reinforce each other to produce and reproduce women's sexualities and genders as different than and distinct from men's sexualities and genders. . . . We resist the tendency to rely on queerness as a unifying category of analysis, keeping the theoretical lens on women and those with female bodies as a more productive angle to reveal the way meanings are attached to particular bodies. (Blackwood et al. 2007, 12)

Their categories of "female" and "women" are thus both situated at the *polar* opposite of "men," and these terms subsume all other identities, including "transgendered females" and "masculine females." In their introduction, Blackwood and Wieringa claim, "In the 1970s and 1980s European and American lesbians and transgendered men were proudly reclaimed as 'women' who resisted patriarchy. . . . Butches were made over into 'real' women, not imitation men, while transgendered men were said to be 'women' who passed. This categorization assumed that having a female body created a natural link, which was defined as 'woman'" (1999, 19). For them, the anatomically-female body is the natural link that unifies lesbians, butches, and transgendered men, and all identities are subsumed under the category of "women." Their interpretation of European and American lesbian and transgender history is also highly suspect given other historical research, in which having an anatomically-female body for transmasculine individuals definitively did *not* provide a "natural link" with other anatomically-female individuals, nor was there an identificatory affinity with the definition of "women" (Meyerowitz 2002; Prosser 1998; Stryker and Whittle 2006).

Only two genders exist in Blackwood and Wieringa's cosmology and "women's" sexualities and genders are distinct from "men's" sexualities and genders. What is especially troubling is that they subsume all women, "transgendered females," and "masculine females" under the category of women in both Western and non-Western contexts. Given the historical and contemporary power imbalances between Western researchers and non-Western study subjects, it seems particularly egregious that they impose their Euro-Western binary-gendered framework onto non-Western/Asian individuals *despite* how some of these individuals may identify.

For instance, Blackwood and Wieringa concede, "Many masculine females who desire women . . . prefer to be called 'men'" (1999, 6). Blackwood's own ethnographic research on Indonesian tomboys reveals that "tomboys see themselves as men" (1999, 185). And in Megan Sinnott's book-length monograph *Toms and Dees: Transgender Identity and Female Same-Sex Relationships in Thailand,* Sinnott writes that masculine-identified *toms* in Thai society "were understood to be 'like men'" (2004, 53). Even elderly Thais that Sinnott interviewed anecdotally referred to individual *toms* as "like a man" (54) and "the woman who was a man" (56). *Toms* themselves would make such statements as "My mind is a man's. . . . I am like a man who has a woman [partner] and must take responsibility for her, like that" (Sinnott 2004, 85).[4] It would thus appear that these transmasculine Asian individuals actually complicate Blackwood and Wieringa's simplistic distinction between "women's" and "men's" sexualities and genders since they identify as or "like" men, although Blackwood and Wieringa categorize them as "women."

Blackwood and Wieringa furthermore resist regarding transmasculine Asian individuals as a gender separate from anatomically-female women-identified people, although evidence proves otherwise. For instance, Sinnott writes that *dees*, the feminine partners of *toms*, were regarded as "ordinary women" who were not considered homosexual, whereas *toms* were marked with a gender different from "ordinary women" (2004, 81–82):

> Most *toms* positioned themselves as situated between ideal masculinity and femininity, strategically accessing claims to both genders, yet simultaneously distancing themselves from both "men" and "women." The sense of blending of masculine and feminine categories has been inherited from Thai understandings of *kathoeys*, who are often seen as being a blend of both sexes. . . . *Toms* often stated that they were "not women"—that is, that they were of a different nature from their female partners and all other "ordinary women." . . . [C]ontemporary Western discourses of "gay hormones" were also mentioned at times, leading to a partial claim of physical difference from *dees*. (84)

It is apparent from this analysis that *toms* view themselves as a separate "nature" and thus a separate sex/gender from the category of "women." This perception stems from an indigenous Thai sex/gender system but is further reinforced by more modern discourses that explain possible *biological* differences, in order to emphasize a *material* difference between the two categories.[5] This clearly refutes Blackwood and Wieringa's subsumption of all transmasculine identities, including *toms*, under the category of "women" or even "female," for that matter, if *toms* are also claiming physical/material difference. In spite of the insights that Sinnott's research provides, Blackwood and Wieringa only incorporate them insofar as they included a chapter by Sinnott in *Women's Sexualities and Masculinities* (Blackwood et al. 2007).

Refutation of the Third-Sex/Gender Paradigm

In tandem with this binary gender oppositionality, then, is the dismissal of the "third-sex/gender" concept as well as the term "queer." Although the third-sex/gender concept is very well known in the study of non-Western subjects, especially in the field of anthropology (Herdt 1996), Blackwood and Wieringa repudiate that:

> The usefulness of the concept "third-sex/gender" seems to lie more in its destabilizing of the dominant bipolar model of gender than in its analytical applicability. . . . Further, an undifferentiated "third-sex/gender" category does not take into account the particular system of women's subordination in the culture concerned. In that sense this theory may depoliticize and ultimately deny or even legitimate the oppression persons with a female body face. Thus we feel the concept is too analytically unstable and too insensitive to gender subordination to be applicable to women's erotic friendships or transgendered females. (1999, 24)

As justification for their repudiation, Blackwood and Wieringa state that both Jakartan transmasculine "butches" and Indonesian transfeminine *waria* refer to themselves as "third sex/gender" (1999, 24), but they don't belong to the same third-sex/gender category.[6] Then, Blackwood and Wieringa state, "Butches seem to have a harder time than the *waria* have. The butches are socially less visible, and their transgression of accepted gender borders is less accepted.[7] As a result they face severe harassment and are socially and economically marginalized" (24). Comparing two marginalized groups necessitates a great deal of evidence and sophisticated analysis to support any observed differences; feminist scholars and writers also critique the project of hierarchizing oppressions (Lorde 1983; Martinez 1993). In Blackwood and Wieringa's attempts to hierarchize the oppression of "butches" as greater than *waria,* they do not present any evidence, nor do they seem to consider socially marginalized aspects of the lives of many *waria,* which have led to a high percentage of participation in survival sex work and HIV seroprevalence among this population; *waria* are often considered to have the highest HIV rates among all gender and sexual minority populations in Indonesia (Joesoef et al. 2003; Morin and Butt 2009; Pisani et al. 2004).

It appears that Blackwood and Wieringa can only consider "women's subordination," and it is used to justify the compression of a variety of anatomically-female and transmasculine individuals into the "women" side of the binary gender oppositionality. They also do not consider that within a given cultural context, the level of subordination of anatomically-female women-identified individuals may be measured to be vastly different than for transmasculine individuals. Depending on what measures are utilized, this difference in levels may be even more vast than the levels of subordination that are measured between transmasculine and transfeminine individuals. Thus, within their binary gender oppositionality, there is no room for the consideration of the "regulatory practices, state and religious ideologies, and global processes" that result in differential subordinations of multiple genders, including *waria. Waria* become relegated to the realm of "men," in which their marginalization is *always already* considered less than that of anatomically-female transmasculine "butches."

In their dismissal of the third-sex/gender concept, Blackwood and Wieringa again interchange "women" and "female" so that "women's subordination" in one sentence becomes "oppressions of persons with a female body" in the following sentence. Given their own troubling definitions of "female" and "women," it is unclear why the third-sex/gender concept is critiqued for being insensitive to gender subordination. (In)sensitivity to gender subordination would depend on how this third-sex/gender term was deployed and applied, and if such studies engaged with questions of gender subordination in their data collection and analysis.

In fact, the terms Blackwood and Wieringa utilize to mark gender subordination, such as "female" or "women," do not inherently refer to other types of inequities and subordinations, for example, racial, economic, and geopolitical inequities, although these gendered terms can still be useful categories of inquiry and analysis. Generalized statements regarding the primacy of "women's oppression," presented with no evidence, universalize an abstract "women's oppression." The overdetermination of this type of oppression occludes other forms of oppression and probably does not reflect the actual social context being studied. It is precisely because these terms have often been deployed (most often by Euro-Western feminist researchers) *only* in the analysis of gender, without examining more complex intersections and imbrications of multiple forms of oppressions, that scholars such as Chandra T. Mohanty have invoked other terms as categories of analysis. For instance, Mohanty developed the term "third world women," which designates a *political* constituency, not a biological or even a sociological constituency (1991, 7). Qualitatively different than Blackwood and Wieringa's definition of "female," Mohanty defines "third world women" as a constituency-as-alliance that emerges out of a *common context of struggle in political relation to sexist, racist, and imperialist structures.*

As further proof of the inappropriateness of the third-sex/gender category, Blackwood and Wieringa then state that a "multiplicity of genders are not represented in the model, for instance, the femme partners of butches" (1999, 24). By inference, it appears that they are discussing Indonesian and other non-Western "butches" since this statement follows a discussion of Indonesian "butches" and *waria.* To add to this vagueness, in "Globalization, Sexuality, and Silences," Blackwood and Wieringa (2007) also admit that they use the term "butch/femme" as a general denominator for a "masculine female" and her partner because they assume that the indigenous terms used by the Asian individuals would cause "confusion" for their readers. Blackwood and Wieringa also admit that "butch/femme" is an Anglo-American term that they are applying to non-Western people (2007, 18). It is thus disturbing why this Euro-Western term is being applied to all non-Western/Asian couples involving a transmasculine person and a feminine partner, for even a cursory glance at descriptions of such Asian couplings often reveal identities and practices that are far different from how Euro-Western butch/femme couples have been described in comprehensive studies (Faderman 1991; Kennedy and Davis 1994).[8]

Sinnott's research (2004) on *dees* and *toms*, conducted in Bangkok, Chonburi Province, and Chiang Mai Province in Thailand, and even Blackwood's research on the girlfriends of transmasculine *tomboys* (2007) in Padang, Indonesia, and on the feminine *cewek* individuals who partner with transmasculine tomboys/*cowok* people (1999) in West Sumatra, Indonesia, reveal that the feminine partners see themselves as and are perceived as "ordinary women" or just "women." As aforementioned, *dees* are viewed as ordinary women, similarly to Indonesian "girlfriends" and *ceweks.* "[*Tombois*] describe their girlfriends as normal women.... Tombois' girlfriends see themselves as normative women who happen to be lesbi at this point because their boyfriends are female. Jeni . . . explained to me, 'I am the same as other women'" (Blackwood 2007, 184). "Where tomboys are clearly marked linguistically in West Sumatra, their sexual partners or lovers have no distinct designation or identity but belong to the category woman" (Blackwood 1999, 187).

Since the feminine partners already belong to the gender category of "women," it is thus perplexing why Blackwood and Wieringa would refute the third-sex/gender category because feminine partners were not included. Furthermore, their investment in the use of the "butch/femme" terminology as an overarching framework belies their Euro-Western underpinnings, in which Euro-Western gender and sexual values, identities, and practices are imposed onto non-Western/Asian individuals and communities. In fact, perusing well-known Euro-Western texts on femme identity indicates that Euro-Western femmes explicitly *do not* see themselves in the same category as heterosexual women (Harris and Crocker 1997; Nestle 1992), unlike *dees*, girlfriends, and *ceweks* in Asia. Lillian Faderman also quotes Laurajean Ermayne from the late 1940s U.S. lesbian magazine *Vice Versa,* who discusses the qualitative differences between Euro-Western femmes and heterosexual women, in which femmes are described as "more intensely womanly than jam [i.e., heterosexual] girls" (1991, 171). Femme lesbians are often characterized as integrating *both* masculinity and femininity, although their femininity is more overtly expressed. Because of this gender integration, femme lesbians are also considered to possess more gender capital than heterosexual women and are also sometimes viewed as "non-women" whose gender-variance is cloaked under a veneer of femininity (Faderman 1991; Mandelbaum and Nuccitelli 2001; Nestle 1992; Stevenson and Goldberg 2002).

Thus, a much more apt frame of reference for non-Western/Asian couples would actually be Faderman's concept of *heterogenderal* couplings (1991, 178), a neutral and encompassing term that merely refers to couples comprised of two different or contrasting genders. The deployment of the "heterogenderal" term to refer to non-Western/Asian couples in which one partner is masculine and the other is feminine is much more appropriate than describing these couples as "butch/femme," especially since at least some of the non-Western/Asian identities and practices appear to diverge quite dramatically from Euro-Western butch/femme identities and behaviors that have been described as standard practices in Western culture.

Besides the characterization of non-Western/Asian feminine partners as ordinary women, another divergence is that feminine partners are frequently conditionally bisexual, which is often accepted by the transmasculine partners. For instance, Sinnott describes a practice of *toms* "releasing" their *dee* partners so that the *dees* can partner with men: "The main source of the suffering of *toms,* according to both *toms* and *dees,* was the supposed tendency of the *dees* to break off relationships with *toms* in order to be with men. . . . Chang, a *dee*, agreed. . . . 'I think there are so many *dees* now who will change and have [anatomically-male] boyfriends.' . . . *Toms* often emphasized their partial male status by renouncing claims to women as long-term partners. These *toms* said they recognized that they were not really men and therefore were not suitable for 'normal women' as long-term partners" (2004, 98–99).

Blackwood's own research reveals this: "Girlfriends I interviewed all averred that they expected they would get married at some point to a man, especially if their current relationships [with *tombois*] fell apart. . . . Dedi said, 'If my girlfriend wants to get married, I'll let her. I won't prevent it.' *Tombois* expected that their girlfriends would want children, which was one thing, Dedi noted, that s/he couldn't give h/er girlfriend" (2007, 189). An earlier study by Blackwood indicates a similar pattern: "Dayan said hir [*sic*] partners have always been feminine (proper women) and are really bisexual. 'Unfortunately, they will leave you for a man if one comes along they like,' s/he said. 'It's our fate that we love women who leave us.' Agus's former lovers are now married with children, and some of Dayan's now have [anatomically-male] boyfriends" (1999, 187).

In characterizations of Euro-Western butch/femme couples, there does not seem to be a common behavioral pattern of femme partners first coupling with butch partners and then leaving their butch partners to marry anatomically-male men. On the contrary, it appears that in Euro-Western cultures, a common behavioral pattern was for femme partners to successively couple with butch partners, in which some of these pairings were long-term partnerships (Faderman 1991; also see "Committed Butch-Fem Relationships" in Kennedy and Davis 1993, 278–322). Therefore, in non-Western/Asian couples it appears that the conditional bisexual behavior of many of the feminine partners and the willing acceptance of this behavior by many of their masculine partners are other features that further distinguish these couples from Euro-Western butch/femme couples.

Instead of compressing all anatomically-female and transmasculine individuals into the same biologically essentialist categories of "female" and "women," it may be useful to consider another paradigm, which is the category of the "fourth-sex/gender" (Lang 1997; Roscoe 1998). Through extensive historical analysis of Native American tribes and cultures in North America, Will Roscoe discovered "alternative gender roles" for both anatomically-male and anatomically-female individuals that have been documented "in every region of the continent, in every kind of society, and among speakers of every major language group" (1998, 7). Roscoe's research, similar to research on non-Western people in other cultures and from other parts of the world (Herdt 1996), also characterizes anatomically-male individuals who occupy these alternative gender roles (e.g. "man-woman" or "half-man, half-woman") as third-sex/gender, whereas anatomically-female individuals (e.g. "woman-man" or "half-woman, half-man") are characterized as fourth-sex/gender (1998, 129; 177). Sabine Lang refers to the fourth-sex/gender as "two-spirit/manly females" (1997, 103). Albeit both Roscoe and Lang focus on Native American people, Roscoe does admit that Native American third- and fourth-sexes/genders have parallel "counterparts around the world" (1998, 20).

What is perhaps most insightful is Roscoe's analysis that a multiple gender paradigm recognizes the arbitrariness of sex itself and that both gender *and* sex are social constructions. Sex is not presupposed to be "natural" in its forms, nor is it a "transcendental signified" for defining gender (Roscoe 1998, 127). Roscoe views sex as merely a category of bodies and gender as a category of persons, in which sex constitutes part of the criteria for being recognized as human. Physical differences are, therefore, unfixed or insufficient on their own to establish gender and are less important than factors such as occupational preference, behavior and temperament, and religious experiences. Gender identities are "total social phenomena" (Roscoe 1998, 127), not merely reiterations of sex. Citing Judith Butler (1993), Roscoe also asserts that cultural interventions serve to "materialize" the body *and* sex (1998, 135), and anatomical differences are not necessarily perceived the same within various cultural contexts (see figure 9.1).

Anatomy (X) social roles/cultural interventions —materialize→ sex/body

FIGURE 9.1 Social construction of sex

Roscoe's astute investigations of the social constructions of sex, then, also highlight the reductive assumptions of Blackwood and Wieringa's definition of anatomical-femaleness as *the* defining category. Blackwood and Wieringa do not consider how "female" is socially constructed within the particular cultures they examine. Compounding this oversight is the metonymic substitution of the gender category of "women" with the sex category of "female." Finally, given the extensive evidence that social processes and cultural interventions give rise to sex and gender, a four-sex/gender paradigm cannot be merely mathematically reduced to a binary gender system. As total social phenomena, each of the four sexes/genders fulfills different roles and reflects an inherent potential of gendered divisions of labor within specific societies (Roscoe 1998, 130).

The Reification of "Lesbianism"

In addition to the metonymic merging of the sex and gender categories of "female" and "women," Blackwood and Wieringa also conflate another category with these two categories, which is the category of "lesbian." In the preface, in *Female Desires*, Blackwood and Wieringa write, "As editors, our desire is to put an end to the so-called invisibility of lesbians across the world and to fill the gap in cross-cultural studies of 'homosexuality' by providing an anthology devoted to the exploration of female same-sex relationships and transgender practices" (1999, ix). It appears that they thus begin their volume by transposing "lesbians across the world" with *all* anatomically-female same-sex relationships and transgender practices.

However, in another series of contradictions, Blackwood and Wieringa then state, "We do not mean to highlight the biological aspects of desires but to raise questions about identities and gender that are too easily subsumed under the categories of woman or lesbian," (1999, ix) and "There are also many transgendered females and women, both cross-culturally and historically, whose lives do not fit within the narrow definition of lesbianism" (x). In their introduction Blackwood and Wieringa write, "Further, adopting a 'lesbian' identity, and the naturalizing discourse associated with it by many gay and lesbian activists in Europe and America . . . carries with it the risk of essentializing" (15). It now seems that they realize that there may be, within the scope of their volume, many gender identities that exceed the categories of "woman" or "lesbian."

Yet, Blackwood and Wieringa state in the preface that their first edited volume "situate[s] work in lesbian studies . . . within the theoretical debates that have been at the forefront in studies of sexuality" (x). In the introduction, Blackwood and Wieringa also write, "Our desire is to reflect the diversity of perspectives in lesbian studies" (3). Thus, similar to the contradictory trajectory in relation to the category of "women," they enact a series of statements that at moments appear to question or complicate categories that they repeatedly invoke. Upon further scrutiny, however, it appears that these statements are actually digressions that belie conceptual foreclosures.

Furthermore, Blackwood and Wieringa appear to have a sizable investment in the reification of the category "lesbian," as if the sole purpose of examining "transgender" identities and practices was to reify "lesbian" subjectivities and the field of lesbian studies, even when transgender studies was already constituted as a valid academic field by the time *Female Desires* was published (Stryker 1998). Blackwood and Wieringa's reification is crystallized in some key statements, such as "lesbian studies in general is not just about 'women' or 'lesbians' since it includes transgendered men (female-to-male transsexuals, female two-spirits, Balkan sworn virgins)" (1999, 7). What is surprising is that they cite *no references,* as if it is a known fact that "lesbian studies" includes the study of transgendered men. In fact, there appears to be copious evidence to the contrary—that the study of transmasculine people often counters or exceeds the limits of "lesbian studies" (Prosser 1998; Stryker 1998; Stryker and Whittle 2006).

What is most troubling is when Blackwood and Wieringa deductively impose this reification of "lesbianism," premised on its merging with "female" and "women," onto non-Western/Asian individuals. For instance, Blackwood and Wieringa state that "such forms have been documented for Thailand, where a proliferation of identities and sexual preferences has sprung up in recent decades, including the '*tom-dee*' lesbian relations" (1999, 19). One of the two references they cite after this statement is an article by Rosalind Morris (1994) on two juxtaposing sex/gender systems in contemporary Thailand—a premodern three-sex system and a modern four-sexualities system. Although Morris does refer to *tom-dee* couplings as lesbian relationships, Morris's deployment of the "lesbian" term is skillfully contextualized within Thai logic, history, and both sex/gender systems and cannot function merely as an abstract reification of Blackwood and Wieringa's category of "lesbian."

In fact, Morris's insightful examination of the premodern Thai system of three sexes and the modern Thai system of four genders implicitly critiques Blackwood and Wieringa's gender frameworks. As opposed to the biologically essentialist definition of "female" as a category of analysis, Morris discusses the premodern Thai tripartite system of male, female, and *kathoey* (transvestite/transsexual/hermaphrodite), in which the third identity of *kathoey* was *biologically irreducible* (1994, 19) and possessed the same ontological status and of equal materiality as the other two identities (21–22). Morris states that the "triadic logic" (20) in Thailand is in contrast to the "primary binarity" in the West (21).

In contrast to Blackwood and Wieringa's use of "genitals" as indicative of a "female" subjectivity and the "materiality" of the female body, Morris writes that "this genital identity is not deemed adequate as a criterion of gendered identity with the system of three. . . . This is a system in which bodies exist less as the locus and origin of consciousness than as an iconic group or symbolic potential, which is realized through everyday practices" (1994, 25). Within this tripartite system, then, female genitals do not necessarily give rise to female materiality, for the symbolic potential indicated by female genitals may also give rise to *kathoey* materiality.

Amalgamating Roscoe's and Morris's concepts, then, produces a *decompressed gender* model of *four sexes,* in which the materiality of sex is realized through social roles and context (see figure 9.2).

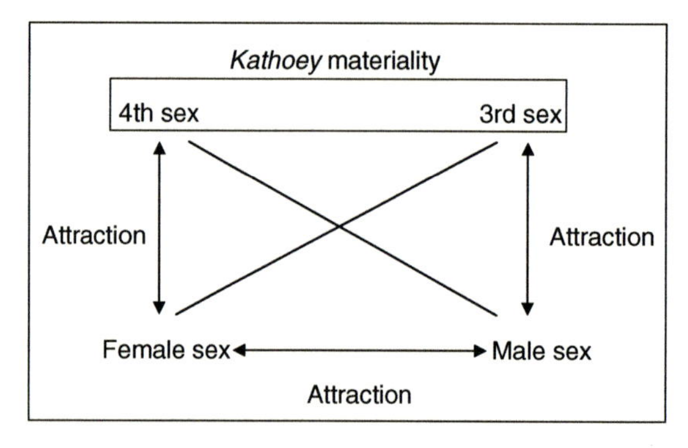

FIGURE 9.2 Decompressed gender model of four sexes

In contrast to the premodern system above, the modern four-sexualities system is informed by the West and is comprised of the opposition between men and women and another opposition between heterosexual and homosexual. Morris repeatedly states that in Thailand this modern system has not replaced the premodern system and that the two occur concurrently (1994, 28).[9]

Of particular interest is Morris's linguistic analysis of "*tom*" and, according to Morris, an assumption among Thais and expatriates that many sex workers are *toms* (1994, 31). Thus "*tom*" connotes both the masculine sexual aggression of a sex worker *and* a stereotypically masculine demeanor, and Morris states that "*tom*" actually represents an "anti-woman" (31), which is in contrast to the category of woman in Thailand. This, of course, refutes Blackwood and Wieringa's categorization of *toms* as "women."[10]

In addition, Blackwood and Wieringa's ascription of *tom-dee* relationships as "lesbian" does not take into account that many *toms* themselves do not identify as lesbian, and at times this disidentification is vehemently stated. According to Sinnott,

> *Toms* and *dees* both generally disdain the term "lesbian" and its sexual and homosexual connotations: "I am a *tom*. I am not a lesbian. I feel disgusted when I hear that word. It isn't good at all; I don't like it at all. Whoever hears it probably won't like it. Somebody want to be like that? It [being a lesbian] isn't the same at all [as being a *tom*]." . . . Another *tom* expressed disgust at the sexual connotations of the term "lesbian": "Lesbian, huh? It's disgusting . . . like 'hysterical' people. . . . I am a *tom*, and my partner is a *dee*." . . . A *tom* quoted in a Thai magazine strongly opposed the term "lesbian" as a self-referent because it neglects gendered distinctions between partners: "*Toms* aren't the same as lesbians. *Toms* are men and [perform sexually] for women. . . . People like to think *toms* are lesbians, and it makes me angry. If somebody says this, I'll punch them. It's an insult." (2004, 30)

In light of such fervent opposition, it is disconcerting why Blackwood and Wieringa would deductively impose the category of lesbian onto non-Western/Asian individuals who so clearly appear to resist this categorization.

Because their interpretation of the category lesbian is premised upon a binary gender opposition, it seems predictable that Blackwood and Wieringa would dispute the use of the category of "queer," similar and parallel to their refusal of the "third-sex/gender" category. They also employ similar logic by stating that "queer" is not sensitive to class and race differences (1999, 21); although, as aforementioned, the categories of female, woman, and lesbian are also not inherently sensitive to class and race differences. In "Globalization, Sexuality, and Silences," Blackwood and Wieringa write, "We argued against the use of the term 'queer' . . . [since] the homogenizing 'queer' subsumes the marked social, economic, and sexual forms of oppression lesbians face in contradistinction to gay men" (2007, 2). Yet they assume the only forms of oppression lesbians face are solely or primarily in contradistinction to gay men, that is, primarily based on sexual difference. However, evidence to the contrary could also be provided; for instance, a group of lesbians may experience great variation of severity and duration of social, economic, and sexual forms of oppression *among* these lesbians; this variation of severity and duration of oppression may be even greater than comparing a specific group of lesbians with a specific group of gay men.

In another gesture of reification, Blackwood and Wieringa state, "We think it is critical at this point in lesbian studies to refuse the ungendered category, whether 'queer' or 'homosexual,' while at the same time refusing reductive definitions or artificial boundaries for lesbian studies. Within the U.S. there is a movement by some lesbian scholars to maintain the visibility of the term" (1999, 21). While there may be some lesbian scholars in the United States who have mobilized to maintain the visibility of the "lesbian" term, it is unclear what this U.S. mobilization has to do with the study of non-Western/Asian individuals. Blackwood and Wieringa appear to be so engaged with this *U.S.* mobilization and apparent struggle for "lesbian" visibility that they do not discern the relevance of this mobilization, if any, to the non-Western/Asian individuals they study. If this mobilization is actually only relevant to Western individuals such as Blackwood and Wieringa, then the study of non-Western/Asian individuals is mere fodder for the reification of their position and investment within this U.S.-based mobilization and struggle. Blackwood and Wieringa then state, "Despite the problems of misrecognition and ethnocentrism that use of the term 'lesbian' occasions, this term needs to remain visible for Western and non-Western audiences alike. Studies of 'lesbians' have for too long been submerged in study of homosexuality in the West. 'Lesbian' remains the sole signifier that distinguishes female/women's same-sex eroticism from men's. 'Lesbian' carries political meaning because it demands recognition for women's differences" (1999, 21).

It is ironic that Blackwood and Wieringa critique the category "queer" for its insensitivity to class and race differences, yet advocate for the category "lesbian" because of its recognition of women's differences, even if it may *also* erase class and race differences. What is also disturbing is in stating that "lesbian" is the "sole signifier" that distinguishes anatomically-female/women's same-sex eroticism, they reveal their apparent ignorance or disregard for the category "women who have sex with women" (WSW) that is commonly invoked in the public health field, medical literature, and in activism—most notably among, and referring to, low-income/poverty-class anatomically-female/women/people of color (Arend 2003; Diaz et al. 2001; Hwahng and Lin 2009; Johnson 2007; Ompad et al. in press). WSW as a category was constructed *precisely* because of the diversity of anatomically-female/women's same-sex eroticism in both

Western and non-Western contexts that could not be captured or defined by the categories of "lesbian" or "bisexual woman." Especially among low-income/poverty-class anatomically-female/women/people of color, there is often little concordance between sexual identity and sexual behavior (Scheer et al. 2003). Blackwood and Wieringa's myopic reification of lesbianism thus occludes categories such as "WSW," and they do not delineate how lesbian became the sole signifier when other signifiers such as WSW are commonly used and deployed. And what exactly is the political meaning of lesbian if WSW was invented precisely because the definitive scope of lesbian was recognized as limited?

Of course, the category of WSW would have to be considerably unpacked in order to ascertain its applicability to the non-Western/Asian individuals that Blackwood and Wieringa study. However, the mere existence and common usage of WSW, *even within Western/U.S. contexts,* indicate the limitations of the category "lesbian."

Conclusion

Blackwood and Wieringa operate through a framework of biological essentialism and binary gender oppositionality, in which "female," "woman," and "lesbian" are metonymically interchangeable on one side of the binary. They also extend the project of heteropatriarchal capitalism and Western neocolonialism through a series of impositions of these Western-informed definitions and categories onto non-Western/Asian individuals. What is confusing is that Blackwood and Wieringa also enact a series of digressions that appear to complicate their gender categories, yet a careful line-by-line analysis reveals how these supposed interrogations toward more complexity are quickly foreclosed with their rather reductive frameworks.

Because Blackwood and Wieringa occupy positions of much more status and socioeconomic privilege than the non-Western/Asian individuals they study, their impositions occur through power imbalances and, thus, must be construed as forms of ideological domination, in which non-Western/Asian individuals become fodder for an agenda that may have more relevance for Blackwood and Wieringa than for the individuals they study. In addition, evidence indicates that for the non-Western individuals they study, social context is fundamental to the materialization of sex, yet they focus on non-Western/Asian anatomically-female bodies, devoid of social context, as generative of sex and, therefore, gender. Blackwood and Wieringa thus effectively re-objectify these non-Western bodies within a fetishistic gaze that both idealizes these "gender ambiguous" individuals and views them as exotic curiosities that can help fulfill their mission to rectify the "invisibility of lesbians across the world." They somehow seem to overlook the fact that the transmasculine Asian individuals themselves do not identify as lesbians and are often extremely opposed to this identification. And it is perhaps not so much lesbians that are invisible, but a diversity of anatomically-female and transmasculine identities and practices, so that even researchers in fields with traditionally underdeveloped gender paradigms, such as public health, realize the limitations of "lesbian" and will instead invoke "WSW."

It is thus only through the examination of sex and gender as total social phenomena that culturally relevant linkages can be revealed. For example, the "*tom*" reflects an inherent potential of one type of role in Thai society, and the "sex worker" reflects an inherent potential of another type of role (Morris 1994). That both roles coalesce under the sign of "anti-woman" indicates complicated intersections and imbrications between the two roles that may reflect overlapping, inherent potentials of masculinized, gendered divisions of sexual labor. An excavation of the intersections and imbrications between these two roles would not only clarify understanding of Thai society but also contribute to the understanding of Asian heteropatriarchal oppressions that may have historically and coercively transformed fourth-sex social roles from that of community or spiritual leadership into sex workers *par excellence,* whereby the fourth-sex is "redeemed" within the system of heteropatriarchal consumption (Sinnott 2004; Sturdevant and Stoltzfus 1993).[11] This may have resulted in an indigenous Asian paradigm that links masculinity with anatomical-femaleness as a site of denigrated sexual labor; this site has historically been applied to anatomical-females of an entire nation within colonial enterprises, such as the defeminization and bastardization of Korean sex slaves by the Japanese military during the Pacific War (Choi 1997; Hicks 1995; Hwahng 2004, 2009; Schmidt 2000).

As Morris states, "Perhaps, in the end, what Thailand tells us is that the essentialism of the two [sexes] can only be overcome with an essentialism of three—or more. Perhaps it tells us that the limits of our genders are simply the limits of our language. It most certainly tells us that heterogeneity is the point at which analysis either achieves lucidity or becomes an agent of occlusion and domination" (1994, 39).

ACKNOWLEDGMENTS

I would like to thank Jian Chen, Kendra Moore, and Amy Sueyoshi for their comments and feedback. I especially thank Jill Fisher for her patience, guidance, and feedback.

NOTES

1. In reviewing Blackwood and Wieringa's research and scholarship, a careful and intensive analysis is necessary. Because of space constraints, this chapter will primarily focus on the conceptual and theoretical frameworks operating within their research methodologies, although other aspects of their methodologies also merit scrutiny.

2. Halberstam (1998) also engages in questionable deductive reasoning in regards to the concept "female masculinity," in which "female" appears to be naturalized as materialized sex, and "female," "woman," and "lesbian" are often conflated together. Halberstam attempts to proffer an abstract universal notion of "female masculinity," which in itself is problematic, especially since this concept is perhaps most applicable within a Euro-Western context.
3. For further explication of why the term "trans/gender-variant" is utilized, see Hwahng and Lin 2009; Lang 1997; Jacobs et al. 1997; Roscoe 1998.
4. Sinnott (2004) states that the phrase "masculine women" to refer to *toms* is actually contradictory, and she avoids using this terminology. Sinnott also finds "female" problematic (2004, 219).
5. A sex/gender system refers to the set of arrangements within historical and cultural social mechanisms by which gender and sexuality are produced and in which a society transforms sexuality into social roles (Rubin 1975).
6. Other evidence indicates that transmasculine Indonesians often use other anglicized terms such as "andro" rather than "butch" (Dania, personal communication, New York City, 2010).
7. Blackwood and Wieringa operate on an assumption that visibility is automatically reflective of social agency and political power. Peggy Phelan astutely observes that "if representational visibility equals power, then almost-naked young white women should be running Western culture. The ubiquity of their image, however, has hardly brought them political or economic power" (1993, 10). *Waria* may, indeed, be more visible in certain social realms, but this visibility does not necessarily translate into less marginalization or more access to social resources.
8. Although Kennedy and Davis also interviewed African Americans and Native Americans, the sample sizes of both of these groups were so small that they admit that their findings were inconclusive (1993, 23).
9. Other non-Western cultures negotiate concurrent indigenous and Western sex/gender systems by deploying the term "homosexual" as a term for the third- or fourth-sex, with heterosexual-identified men or women often participating in bisexual behavior (Johnson 1997; Kulick 1998; Prieur 1998).
10. It is important to view the modern Thai four-sexualities sex/gender system as different from Roscoe's (1998) four-sex/gender system. Each of the four genders fulfilled specific social roles in Native American societies, but within modern Thai capitalist heteropatriarchy, the non-sex-worker *tom* has no sustainable exchange value in the political and representational economies of heterosexuality (Morris 1994, 31). *Dees*, on the other hand, are redeemed by enacting aesthetics of femininity and eventually partnering with men.
11. For further discussion of heteropatriarchy, see Smith 2006.

REFERENCES

Arend, E. D. 2003. The politics of invisibility: HIV-positive women who have sex with women and their struggle for support. *Journal of the Association of Nurses in AIDS Care* 14 (6): 37–47.

Blackwood, E. 1999. *Tombois* in West Sumatra: Constructing masculinity and erotic desire. In *Female desires: Same sex relations and transgender practices across cultures*, edited by E. Blackwood and S. Wieringa. New York: Columbia University Press.

——. 2007. Transnational sexualities in one place: Indonesian readings. In *Women's sexualities and masculinities in a globalizing Asia,* edited by E. Blackwood and S. Wieringa. New York: Palgrave MacMillan.

Blackwood, E., and S. Wieringa, eds. 1999. *Female desires: Same sex relations and transgender practices across cultures.* New York: Columbia University Press.

——. 2007. Globalization, sexuality, and silences: Women's sexualities and masculinities in an Asian context. In *Women's sexualities and masculinities in a globalizing Asia,* edited by E. Blackwood, S. Wieringa, and A. Bhaiya. New York: Palgrave Macmillan.

Blackwood, E., S. Wieringa, and A. Bhaiya, eds. 2007. *Women's sexualities and masculinities in a globalizing Asia.* New York: Palgrave Macmillan.

Brown, L. 2001. *STI/HIV: Sex work in Asia.* Manila, Philippines: World Health Organization, Regional Office for the Western Pacific.

Butler, J. 1993. *Bodies that matter: On the discursive limits of "sex."* New York: Routledge.

Choi, C. 1997. Introduction. *Positions: East Asia Cultures Critique,* Special Issue: *The Comfort Women: Colonialism, War and Sex* 5 (1): v–xiv.

Diaz, T., D. Vlahov, B. Greenberg, Y. Cuevas, and R. Garfein. 2001. Sexual orientation and HIV infection prevalence among young Latino injection drug users in Harlem. *Journal of Women's Health and Gender-Based Medicine* 10 (4): 371–380.

Faderman, L. 1991. *Odd girls and twilights lovers: A history of lesbian life in twentieth-century America.* New York: Penguin Books.

Glaser, B. 1978. *Theoretical sensitivity: Advances in the methodology of grounded theory.* Mill Valley, CA: Sociology Press.

——. 1992. *Basics of grounded theory analysis: Emergence versus forcing.* Mill Valley, CA: Sociology Press.

Halberstam, J. 1998. *Female masculinity.* Durham, NC: Duke University Press.

Harris, L., and E. Crocker, eds. 1997. *Femme: Feminists, lesbians, and bad girls.* New York: Routledge.

Herdt, G. 1996. *Third sex, third gender: Beyond sexual dimorphism in culture and history.* New York: Zone Books.

Hicks, G. 1995. *The comfort women: Japan's brutal regime of enforced prostitution in the Second World War.* New York: W. W. Norton.

Hwahng (Wahng), S. J. 2004. The illogics of masculine deterritorialization: Asian and Asian American racial performativities, regendered embodiments, and collective assemblages of enunciation. PhD diss., New York University.

——. 2009. Vaccination, quarantine, and hygiene: Korean sex slaves and no. 606 injections during the Pacific War. *Substance Use and Misuse* 44 (12): 1768–1802.

Hwahng, S. J., and A. J. Lin. 2009. The health of lesbian, gay, bisexual, transgender, queer, and questioning people. In *Asian American communities and health: Context, research, policy, and action,* edited by C. Trinh-Shevrin, N. Islam, and M. Rey, 226–282. San Francisco: Jossey-Bass.

Jacobs, S.-E., W. Thomas, and S. Lang, eds. 1997. *Two-spirit people: Native American gender identity, sexuality, and spirituality.* Urbana: University of Illinois Press.

Joesoef, M. R., M. Gultom, I. D. Irana, J. S. Lewis, J. S. Moran, T. Muhaimin, et al. 2003. High rates of sexually transmitted diseases among male transvestites in Jakarta, Indonesia. *International Journal of STD and AIDS* 14 (9): 609–613.

Johnson, C. A. 2007. *Off the map: How HIV/AIDS programming is failing same-sex practicing people in Africa.* New York: International Gay and Lesbian Human Rights Commission.

Johnson, M. 1997. *Beauty and power: Transgendering and cultural transformation in the southern Philippines.* New York: Berg.

Kennedy, E. L., and M. D. Davis. 1994. *Boots of leather, slippers of gold: The history of a lesbian community*. New York: Penguin Books.

Kulick, D. 1998. *Travesti: Sex, gender, and culture among Brazilian transgendered prostitutes*. Chicago: University of Chicago Press.

Lang, S. 1997. Various kinds of two-spirit people: Gender variance and homosexuality in Native American communities. In *Two-spirit people: Native American gender identity, sexuality, and spirituality*, edited by S.-E. Jacobs, W. Thomas, and S. Lang, 100–118. Urbana: University of Illinois Press.

Lorde, A. 1983. There is no hierarchy of oppressions. In *Homophobia and education*. New York: Council on Interracial Books for Children.

Mandelbaum, B., and A. Nuccitelli. 2001. Fabulosity: The fem warrior princess workshop. Paper presented at the Fifth Annual True Spirit Conference: Celebrating Human Diversity, February 16–19, Washington, DC.

Martinez, E. 1993. Beyond black/white: The racisms of our times. *Social Justice* 20 (1–2): 1–8.

Meyerowitz, J. 2002. *How sex changed: A history of transsexuality in the United States*. Cambridge, MA: Harvard University Press.

Mohanty, C. T. 1991. Introduction: Cartographies of Struggle. In *Third World Women and the Politics of Feminism*, edited by C. T. Mohanty, A. Russo, and L. Torres, 1–47. Bloomington: Indiana University Press.

Morin, J., and L. Butt. 2008. Papuan waria and HIV risk. *Inside Indonesia* 94 (October–December): 1–3. http://insideindonesia.org/content/view/1141/47/ (accessed July 15, 2009).

Morris, R. C. 1994. Three sexes and four sexualities: Redressing the discourses on gender and sexuality in contemporary Thailand. *positions* 2 (1): 15–43.

Nestle, J., ed. 1992. *The persistent desire: A femme-butch reader*. New York: Alyson Books.

Oetomo, D. 2000. Masculinity in Indonesia: Genders, sexualities, and identities in a changing society. In *Framing the sexual subject: The politics of gender, sexuality, and power*, edited by R. Parker, R. M. Barbosa, and P. Aggleton, 46–59. Los Angeles: University of California Press.

Ompad, D., S. Friedman, S. J. Hwahng, V. Nandi, C. Fuller, and D. Vlahov. In press. Risk behaviors among young drug using women who have sex with women in New York City. *Substance Use and Misuse*.

Phelan, P. 1993. *Unmarked: The politics of performance*. New York: Routledge.

Pisani, E., P. Girault, M. Gultom, N. Sukartini, J. Kumalawati, S. Jazan, et al. (2004). HIV, syphilis infection, and sexual practices among transgenders, male sex workers, and other men who have sex with men in Jakarta, Indonesia. *Sexually Transmitted Infections* 80 (6): 536–540.

Prieur, A. 1998. *Mema's house, Mexico City: On transvestites, queens, and machos*. Chicago: University of Chicago Press.

Prosser, J. 1998. *Second skins: The body narratives of transsexuality*. New York: Columbia University Press.

Roscoe, W. 1998. *Changing ones: Third and fourth genders in Native North America*. New York: St. Martin's Press.

Rubin, G. 1975. The traffic in women: Notes on the "political economy" of sex. In *Toward an anthropology of women*, edited by R. Reiter. New York: Monthly Review Press.

Scheer, S., C. A. Parks, W. McFarland, K. Page-Shafer, V. Delgado, J. D. Ruiz, et al. 2003. Self-reported sexual identity, sexual behaviors, and health risks: Examples from a population-based survey of young women. *Journal of Lesbian Studies* 7 (1): 69–83.

Schmidt, D. A. 2000. *Ianfu—The comfort women of the Japanese Imperial Army of the Pacific War*. Lewiston, NY: Edwin Mellen Press.

Sinnott, M. J. 2004. *Toms and dees: Transgender identity and female same-sex relationships in Thailand.* Honolulu: University of Hawai'i Press.

Smith, A. 2006. Heteropatriarchy and the three pillars of white supremacy: Rethinking women of color organizing. In *Color of violence: The INCITE! anthology*, edited by INCITE: Women of Color Against Violence, 66–73. Boston: South End Press.

Stevenson, J., and S. T. Goldberg. 2002. Girls in the boys' room: A discussion on contemporary femme gender. Paper presented at the Sixth Annual True Spirit Conference: Letting Our True Spirits Soar, February 15–18. Washington, DC.

Stryker, S. 1998. The transgender issue: An introduction. *GLQ: A Journal of Lesbian and Gay Studies* 4 (2): 145–158.

Stryker, S., and S. Whittle, eds. 2006. *The transgender studies reader.* New York: Routledge.

Sturdevant, S. P., and B. Stoltzfus, eds. 1993. *Let the good times roll: Prostitution and the U.S. military in Asia.* New York: New Press.

PART FOUR

Medical Interventions

10

Facial Feminization and the Theory of Facial Sex Difference

The Medical Transformation of Elective Intervention to Necessary Repair

HEATHER LAINE TALLEY

Facial Feminization Defined

In the 1980s and 1990s, a Northern California plastic surgeon with extensive experience in reconstructive surgical techniques developed a unique collection of procedures marketed to male-to-female (MTF) transsexuals for the purposes of changing facial appearance.[1] In the United States, facial feminization surgery (FFS) is promoted and practiced as a distinct set of procedures by primarily four surgeons.[2] What distinguishes facial feminization from cosmetic surgery, more generally, is that FFS is expressly aimed at making the face more feminine, rather than simply more attractive—although some accounts of what constitutes beauty suggest that the most feminine faces are often perceived as the prettiest (Etcoff 1999). Facial feminization is accomplished through a variety of procedures including but not limited to a brow lift, a trachea shave, a jawline reduction, a chin reduction, and a face and neck lift. While surgeons are often vague about how much feminization costs, patients' Web sites report that fees for FFS run between twenty thousand and forty thousand dollars.

Like other forms of cosmetic and reconstructive surgery, the results of facial feminization require intensive surgery often lasting hours, sometimes as many as ten. Because procedures are aimed at radically altering multiple facial features, patient's faces (and skulls) undergo serious surgical manipulation. For example, jawline and chin reduction requires actually breaking or severing the bones of the face with a surgical drill. Facial bones are then resecured with screws, wires, and bone pastes. Brow shave or forehead recontouring involves removing a section of the skull, reshaping it with a device that resembles a dremel drill,[3] and reattaching it to the skull. To reduce the distance between the hairline and the eyebrows, a cut is made along the hairline, a section of skin is removed, and the scalp is pulled forward, bringing the hairline down lower on the forehead. Swelling, bruising, and scarring are common. Some patients experience changes in the face's range of motion, reducing the facility with which one moves one's jaw, for example. Others report a loss in facial sensation, typically a sort of numbness that leaves a face unable to perceive human touch. Like all forms of surgery, infection and death loom as potential side effects. And, of course, the resulting face may not resemble what one had desired.

Facial feminization is an invasive, expensive, dubiously successful intervention. And yet it remains a highly sought after technology of transgendering for transwomen.[4] By contrast, female-to-male transsexuals achieve masculinization of the face through hormone therapy, thus there is no such comparable set of surgical practices termed facial masculinization surgery. Unlike other academic analyses that query the significance of genital reassignment surgery as a case of the medical production of gender, I take up modes of transitioning that are aimed at the face. Facial feminization is virtually unaccounted for, and yet because of how uniquely visible our faces are relative to other body parts, it follows that facial interventions carry a unique significance. Particularly for transgender people, facial appearance is crucial in navigating everyday life unencumbered by the threat that not passing as one's self-identified gender presents. I begin by describing the processes through which the male-to-female transsexual face is taken apart, both figuratively and literally, through the deployment of an emergent theory of facial sex difference. I argue that surgeons emphasize the differences between the female face and the male face and ultimately bring into formation a way of seeing the face that inspires the literal taking apart of the skull. While surgeons continually differentiate between sexed bodies, FFS is not simply a mode of "doing gender" but rather, as I demonstrate, a technology of repair. Transwomen's faces become defined as disfigured, and facial feminization itself is positioned as a method of reconstructive surgery. In essence, I argue that the processes through which facial sex difference is articulated work as a diagnostic strategy that inspires intervention. I conclude by situating facial feminization in a technoscientific landscape alongside other gendered medical interventions that, while they might be considered elective, come to carry the weight of necessary interventions, and I point to moments in the practice of medicine in which gender is conceptualized more fluidly. In doing so, I hope to demonstrate how medicine might be divested of the weight of requirement.

This work is based on ethnography conducted through attendance at conferences featuring seminars on facial feminization directed toward potential consumers, namely, transwomen in the midst of transitioning, and content analysis of surgeons' Web sites along with printed materials distributed during seminars. Transgender conferences are unique quasi-public spaces.[5] Overwhelmingly, most of the people who attend and many of the people who present seminars self-identify as transgender. In this way, conferences offer attendees opportunities to network with others with shared interests and concerns and to access specialized information. Transgender conferences feature seminars that address a wide range of topics including law, spirituality, mental health, and activism, but a significant number of featured speakers address topics related to body modifications made possible by and forged through medical technologies. Not surprisingly, seminars on hormone therapy and genital reassignment surgery appear throughout conference schedules. In addition, each of the primary U.S.-based facial feminization surgeons appears at transgender conferences to discuss and market facial feminization.[6] Seminars are opportunities for potential patients to meet and consult with surgeons, to learn more about facial feminization techniques, and to view the faces of other transwomen who have undergone FFS. Thus, seminars are simultaneously information sessions, commercial advertisements, and public spectacles.

Sexing the Face

Feminist accounts of cosmetic surgery have pointed to the ways in which cosmetic surgery consumption is gendered. Not only are women more prone than men to pursue particular interventions, but what men and women hope to accomplish via intervention is largely determined by social conventions about men's and women's bodies (Blum 2003; Pitts-Taylor 2007; Sullivan 2004). As Diana Dull and Candace West argue, "This [gender] is the mechanism that allows them to see the pursuit of elective cosmetic surgery as 'normal' and 'natural' for a woman, but not for a man. The accountability of persons to particular sex categories provides for their seeing women as 'objectively' needing repair and men as 'hardly ever' requiring it" (2002, 137).

In this way, cosmetic surgery is both an effect of gendered relations and a means of doing gender. Cosmetic surgery consumers employ surgery as a means for more closely approximating gendered cultural ideals. Women purchase faces that might be described as beautiful or, perhaps, sexy. Men consume interventions that result in good-looking, though certainly still masculine, faces. Yet not all women who get cosmetic surgery change their bodies in precisely the same ways. There are a range of possibilities, a multitude of means for aesthetic enhancement that one can choose toward more desirably doing one's respective gender. For example, a woman may choose any number of surgically constructed noses—the ski slope, the upturned, or the perfectly angular. The desired outcome is gendered, but it is gendered in a very generic sort of way.

I argue that facial feminization is different. It, too, is a method of surgically inscribing gender onto the body, but facial feminization takes up gender as its very object of intervention (and invention). Facial feminization is not a technology simply employed in the process of doing gender. Rather, facial feminization is about surgically constructing gender itself. It is not, as is the case in cosmetic surgery, directed toward helping women achieve a prettier face. Rather, it is the surgery that inscribes a face that is perceived as female at all. To this end, facial feminization relies on extant, essentialized notions about what distinguishes a male face from a female face.

In a 2007 issue of *Clinics in Plastic Surgery*, in an article entitled "Transgender Feminization of the Facial Skeleton," general plastic surgeons were introduced to the idea of facial feminization by a group of Dutch surgeons (Becking et al. 2007). The article begins by defining transsexualism and describing multiple modalities employed in the treatment of gender identity disorder, specifically, genital reassignment surgery and hormone therapy. Yet, the authors contend that these are crucial though insufficient means of addressing gender identity disorder: "For passing in public as a member of the opposite gender, facial features are of utmost importance for the transsexual individual" (Becking et al. 2007, 558). Beyond suggesting that the face should be conceived as a critical site of intervention, the surgeons propose the following: "There is a need for more objective standardization of the differences in the facial features of the two sexes, to facilitate surgical treatment planning and more objectively assess the outcome of the facial surgery on psychosocial functioning and appearance, not only from the perspective of those treating, but also from the patient's own point of view" (563–564).

By calling for "objective standardization," Alfred Becking and colleagues argue that rigorous scientific research aimed at discovering facial sex differences would help in elaborating a basis for surgical practice and a standard by which to judge success. Although there has been a limited amount of the kind of research they propose, there are already circulating "theories" of sex differences in facial appearance.

In a pamphlet distributed by Douglas K. Ousterhout, the American surgeon often credited with developing facial feminization surgery, entitled "Feminization of the Transsexual" and distributed to prospective consumers, techniques of facial feminization are described in detail. However, the pamphlet is not simply a list of surgical procedures often accomplished in the process of feminization.[7] Rather, the pamphlet offers a theory of facial sex difference and a subsequent account of why facial feminization works. In the introductory section, the pamphlet reads:

> There are basic differences between a male and a female skull, differences long appreciated by anthropologists studying skulls but also by artists as well. Females have a more pointed chin, tapered mandible, and less nasal prominence than males. These areas must be modified from those more massive areas on the male. . . . You must change the underlying structures to affect a real change. Changing the shape of the skull will

> markedly assist in changing one from distinctly male to female. . . . REMEMBER: TO APPROXIMATELY FEMINIZE THE FACE, THE SKULL MUST BE APPROPRIATELY REDUCED TO FEMININE SIZE AND PROPORTIONS.

Ousterhout's theory is comprised of two central claims. First, Ousterhout asserts that anthropological evidence suggests that on average, the skulls of men and women differ in both shape and size. His pamphlet follows a well-established pattern of employing scientific research in the service of identifying sex differences through cranial measurements, hormones, and skeleton size and shape (Gould 1981; Oudshoorn 1994; Schiebinger 1986). In this way, Ousterhout offers an objective, "scientific" explanation that accounts for why face work is important to transitioning from male to female. This theory of facial sex difference also posits that changing the skull through surgery will alter the appearance of the face, specifically, the gendered effect of the facial structure. In effect, Ousterhout's account of facial feminization works to position face work as a crucially important intervention for male-to-female transsexuals, at the same time that it advances essential, naturalized accounts of sex/gender differences.

The pamphlet continues by elaborating the differences between male facial features and female facial features, including the brow, forehead, hairline, chin, mandible (jaw), cheeks, lips, neck, and nose. A single image appears in the pamphlet. It is a black-and-white graphic depiction that more closely resembles clip art than a representative depiction of the human skull. Yet, this image is deployed to visually represent dissimilarities between the "male" and "female" face. The pamphlet is a guide for potential patients that identifies the disparity between the two skulls. The image is a reference point that patients can continually consult to confirm the claims made in the text that follows.

Throughout the pamphlet, facial sex differences are identified and surgical techniques for addressing these differences are described and prescribed. In regards to the forehead, the pamphlet emphasizes the "prominence" of male bone structure: "As the male forehead is so different than the female forehead this may be one of the most important areas to modify. Males have brow bossing with a flat area between the right and left areas of bossing while females have a completely convex skull in all planes and markedly less prominence." To deal with the masculine forehead, bone contouring is used to reduce the bossing (or ridge) that appears across the forehead.

The distance between the hairline and the brow line also becomes salient: "In physical anthropology studies, it has been shown that men have a longer distance from the brows to the hairline than do women. . . . A long forehead is generally acceptable for the male but not for the female." The pamphlet suggests that this distance be shortened by way of scalp advancement and in some cases a brow lift.

The chin also works as a crucial mark of gender: "The chin varies markedly between the male and the female. The male chin is generally wide and vertically high while the female chin tends to be more pointed, narrow, and vertically shorter. . . . Thus the chin is an extremely important area in gender recognition." To address this "important" feature, Ousterhout suggests a sliding genioplasty, which involves cutting and removing sections of bone to reduce the "squareness" associated with male chins.

As the preceding excerpts indicate, "Feminization of the Transsexual" follows a particular pattern. In it, each facial feature is dissected for sex differences, and those differences are described in detail. These descriptions are followed by brief descriptions of surgical techniques aimed at repairing, specifically, re-gendering, faces. In the process, differences are reified and surgery becomes the "solution." The underlying theory of facial sex difference articulated in Ousterhout's pamphlet permeates facial feminization seminars, as well.

Facial feminization seminars function to position facial feminization and male-to-female transsexual faces in particular ways. Specifically, seminars work to forward the theory of facial sex difference that transforms FFS from a collection of elective procedures into a technique of bodily repair. The following excerpt from field notes taken during Southern Comfort 2006 describes the format of one seminar.

> Then, the doctor suggests that "anthropological differences" matter, that "anthropologists can tell the difference between male and female." . . . On his PowerPoint presentation he clicks to a graphic of two different skulls and two different faces (one male and one female). He then begins to point out the differences between the two skulls, demonstrating how each feature of the face varies between men and women. . . . "There are basic differences between the male and the female skull." The angle of the mandible is different. The bossing is different. The nose angle is different. The vertical height of the mouth is different. In most of his procedures, the surgeon removes skin to minimize the distance between the hairline and the brows. . . . For cheeks, he suggests cheek implants, which come in different sizes, but he says that not everyone needs them. "The cheeks are not male or female." He also suggests upper lip shortening. The doctor argues that this is necessary because men only show lower teeth when they smile and women show upper teeth. . . . Females, he argues, have a tapered face, and males have a square face. He suggests a sliding genioplasty and a lateral mandibular reduction to eliminate squareness. (field notes)

This theory about facial sex difference serves as the working framework for approaching facial feminization.

Other surgeons, however, take issue with competing surgeons' "theories." For example, each surgeon puts a varying degree of emphasis on the importance of soft tissue work relative to bone work. But every surgeon relies on similar strategies in his seminars. Each reviews images of male and female faces to emphasize facial differences and describes how facial feminization will affect appearance. In practice, seminars resemble each other, despite the particular surgeon who is featured. Consider the seminar structure employed by Dr. Adams:

> To illustrate how much bone work is needed to appear female, the doctor posts two pictures side by side on the screen. A male skull and a female skull. He begins to point out male and female facial characteristics. "When you look at your female counterparts, I'm not trying to be rude, but there are differences." Dr. Adams points out a number of facial features—the temporal ridge, facial hollowing, the cheeks, the eyebrows—and points out the differences between the male skull and the female skull. "Soft tissue is the magic." . . . "Procedures on the upper face are the most feminizing. . . . The forehead is the most critical thing." (field notes)

In this seminar, images of skulls are projected side by side for the audience to see. One by one, each facial feature is examined, and in each case, differences are highlighted. A presentation by Dr. Thomas is virtually indistinguishable:

> The surgeon poses the following question to the audience: "What is it about the face that allows the distinguishing of gender?" He compares slides of men's and women's skulls and argues, "The thing that's really making the difference—the bone. The skin is just the skin." Dr. Thomas goes through parts of the face one by one and describes what is needed to make the face appear more feminine. Using a laser pointer, the surgeon riffs on each face that fills the screen, identifying what features appear too masculine. He tells the audience that he feminizes the face through neck surgery (tracheal shave), forehead remodeling (osteotomy and ostectomy), feminization of the jaw (width reduction), and feminization of the nose. (field notes)

In every seminar, then, surgeons use images to demonstrate and to foster notions of facial sex difference. Selective images are invoked as "proof" that the male face is demonstrably different from the female face. In this way, facial feminization surgeons take a reductionistic approach. Single images of a male skull and a female skull are positioned to represent sex and gender writ large. In this way, surgeons rely on a theory of sexual dimorphism that presumes real, measurable difference between men and women and downplays variability in appearance among the categories of men and women. This strategy works to reify the differences between facial appearance in men and women, and thus to reproduce sex/gender differences. Instead of using images as representations or examples of male and female faces, they are used as evidence about general patterns, which then inform surgical practice and patients' notions both about what is "wrong" with their pre-surgical face and what a new face might look like. This theory of facial sex difference saturates the approach to FFS because even though surgeons are competitors, each employs virtually identical rhetorical strategies and logic. In other words, while each is contending for patients in what is a relatively narrow market, each describes the process in such similar ways that there appears to be no other means for looking at the transwoman's face.

In the talk of FFS surgeons, the female face is deciphered, dismantled, and identified in the service of identifying the truth about what constitutes a feminine face. In the process, a surgical standard is constructed by way of fetishizing facets of femininity, in this case feminine facial features, and positioning these features as constitutive, as opposed to indicative, of femininity. The female face becomes the symbolic standard against which a patient's real face is compared. Facial features that do not correspond are subject to intervention. Those features that do not evoke the female face are deemed masculine. In this way, masculinity becomes an empty signifier, a repository for all features not defined as female. This stands in sharp contrast to a long history wherein female bodies have been conceptualized simply as not male (Tavris 1993).

It is impossible to decipher how "real" facial sex differences are. As feminist science studies scholars suggest, science does not unfold outside of culture (Haraway 2004). Yet in a cultural context in which sexual dimorphism permeates our approaches to bodies, doctors seem to persuasively argue that men's and women's faces are not only fundamentally different from one another but rather relatively homogenous within each sexed category. In fact, one only need look around any public space to see that there is no facial characteristic shared by all women or by all men. At the same time, something about faces makes it possible for us to distinguish gender even without the cultural signifiers—makeup, facial hair, or accessories—in many cases. Though many faces, when taken on their own, are ambiguous. Given the tenuousness of designating difference, coupled with surgeons' insistence that differences are profound, it is entirely possible to imagine that "biological" women might also become subject to FFS for "insufficiently female" faces. What the case of FFS demonstrates so vividly is the way in which sex differences become hyperbolically deployed to inspire intervention.

While surgeons publicly dissect the images and point to differences in almost every facial feature, in the process male and female faces are positioned not simply as dissimilar from one another on average but rather as drastically divergent and thoroughly problematic for transwomen. Dr. Thomas articulates the degree of difference in this way: "It's not just bone work and pull some skin; you need a global change."[8] By "global," the doctor seems to suggest that facial feminization is a surgical overhaul. This radical change is not simply desired, that is, elective, but rather for the transwoman it is "needed." Dr. Adams similarly suggests that the change needed is drastic: "In order to look female you must change your skeletal appearance. To do less is absolutely wrong. Less is not more." To accomplish the objective of facial feminization, that is, looking appropriately female, requires not only *some* surgical intervention, but rather, as the previous excerpt suggests, a drastic reworking of the skeleton. By emphasizing how invasive facial feminization needs to be, surgeons imply that the male face is untenable. It is, in short, a problem, as Dr. Peterson suggests in a seminar: "He turns to pictures of 'real' men and says that the pictures illustrate a 'female face on males.' Dr. Peterson says this is 'not a problem.' Then he shows a picture of a 'male face on a female' and says, 'but a male face on a female really is [a problem].'"

What makes traces of masculinity so unsustainable is not altogether clear in the surgeons' discourse, but the diagnosis is simple enough: if one wishes to live as a woman, then one's face cannot appear masculine. It is unnatural.

The Transwoman's Face as Disfigured

The faces subject to facial feminization are not immediately intelligible as disfigured faces, yet within sites of facial feminization, the faces of male-to-female transsexuals are positioned as untenable, and thus the face is in *need* of repair. Certainly, impetus of repair implicitly mobilizes the specter of disfigurement, but explicit references to disfigurement appear frequently in subtle and not so subtle ways. Faces that might commonly be referred to as ugly are discursively constituted as disfigured via brief, but revealing, references. In one facial feminization seminar, a surgeon pointed to a photograph of a seemingly unremarkable face, remarking, "You can see her forehead deformity." Words like deformity mark the face not simply as masculine but rather as fundamentally disfigured. In another seminar, a doctor explained why somebody might choose facial feminization: "This is the same thing as if you were in a car wreck and you want to look like who you really are." The male-to-female transsexual face is equated here with an injury or a trauma, and thus in need of repair. From this perspective, facial feminization is hardly elective surgery but is rather positioned as unequivocally necessary—especially by those who stand to benefit commercially.

Perhaps even more revealing is a strategy used by one surgeon, Dr. Peterson. The first time I attended a seminar presented by this surgeon, I watched as his assistant connected his laptop to a projector. As the doctor and his assistant readied themselves for the lecture, images flashed on a large screen that appeared at the front of the hotel meeting room. When I saw the photographs, I was sure that his assistant had opened the wrong PowerPoint file. What filled the screen were images of children with a range of cranio-facial anomalies. These images seemed shockingly out of place. Why begin a lecture on facial feminization with the faces of facially variant children?

Field notes from a second seminar presented by the same surgeon reveal why it made precise sense from the perspective of the surgeon to begin with such images:

> Dr. Peterson tells the audience that he spent twenty years running a center focused on the repair of cranio-facial anomalies. He's worked on hard cases, "horrible" facial anomalies. He flips through slides of people with a variety of congenital facial differences. One has a cleft palate. One has an unusual skull shape. It is not spherical. The surgeon refers to the photograph as an example of "clover leaf skull." One picture of a baby illustrates asymmetrical facial features. The eyes and the nose appear randomly placed as if in a Picasso painting. One has eyes that upon profile extend dramatically beyond a recessed eye cavity, resulting in look of extreme surprise. You can see the shape of the entire eyeball, and they look as though they may fall out of the skull. Next to each is an "after" picture that shows the face post-surgery. Each face looks remarkably different, more "normal." Dr. Peterson remarks, "Those three patients show what I do to feminization patients." (field notes)

Without a doubt, the images convey technical skill and surgical "successes," but this introduction accomplishes much more than that. By beginning the presentation with images of children with craniofacial anomalies, the audience is immediately engaged with representations of disfigurement. While looking at these pictures, the audience was solemn. The pictures appear within a cultural lineage in which photographs of disabled children are to be witnessed as evidence of the tragedy of congenital difference. By claiming that "those three patients show what I do to feminization patients," the faces of the transwomen in the audience are positioned as somehow just like the faces of the disfigured children—in need of repair. In addition, such a claim characterizes facial feminization as somehow akin to reconstructive surgery, as opposed to elective cosmetic intervention.

Other surgeons offer a narrative about their own careers that firmly locates their facial feminization practices within a trajectory of reconstructive surgery. A pamphlet distributed by an East Coast surgeon, Dr. Thomas, describes his educational history and professional affiliations: "Advanced training was obtained with fellowship in Facial Plastic and Reconstructive Surgery, and Microsurgery through Harvard Medical School. He currently devotes his practice to facial plastic surgery and head and neck cancer reconstruction. The busiest component of his practice is Facial Feminization Surgery (FFS)."

Another, Dr. Adams, emphasizes similar career ties in a handout distributed to potential patients entitled "Head of Plastic Surgery of Nation's Oldest and Busiest Military Hospital Relocated to Chicago." The feature describes the techniques innovated by the surgeon while serving as the chief of the Plastic Surgery Department at Naval Medical Center, Portsmouth, Virginia. Plastic surgery departments located at military hospitals often function to innovate reconstructive techniques that address disfigurement resulting from war injuries. Thus, work as a plastic surgeon in the military locates one's professional history within the domain of reconstructive surgery. I am not arguing that it is unique for cosmetic surgeons to have training in reconstructive techniques; rather, I am interested in the ways in which surgeons use that training to communicate something about the work they do as facial feminization surgeons. By positioning facial feminization as a logical extension of training accomplished toward the ends of reconstructive surgery, surgeons both produce and rely on associations between disfigurement and the male-to-female transsexual face.

It is not simply that facial feminization and reconstructive surgery are discussed as technical equivalents and that transwomen's faces are characterized as disfigured faces. More significantly, facial feminization surgeons invoke meanings attached to reconstructive surgery to accomplish the work they do to feminize the face. Specifically, facial feminization is described as life-changing work—both for surgeons and patients. Dr. Nelson, a surgeon whose practice is divided between facial feminization and genital reassignment surgery, begins a presentation by explaining to the audience what makes a surgeon trained to "fix" facial "abnormalities" decide to start one of the country's preeminent centers for transgender related surgery: "What deformed people remember is what they looked like before." He continues, by telling the audience that his patients are happy to be in the hospital. His patients are happy to see him.

In contrast to presumably unhappy "deformed" patients, this doctor suggests that the work of facial feminization is about making patients happy. Rather than being preoccupied with their pre-surgical visage, the surgeon suggests that transwomen embrace their interventions and that this gives a particular significance to his work. Dr. Adams goes further. When asked by an audience member, "Why do you do this?" he replies, "I came out of the navy with a great set of tools. . . . As a plastic surgeon, it is rare that I can make a difference. I have profoundly affected their [transwomen patients] life so that they can go in society and live their lives. We all know how cruel society can be."

This surgeon frames facial feminization as life-changing work, as a form of repair that allows people to "live their lives." The focus is on changing the individual rather than the society that presumably makes living life difficult for transwomen.

By introducing facial feminization as an extension of reconstructive surgery, FFS is unequivocally positioned as a kind of repair for disfigurement. Facial feminization surgeons "fix" disfiguring conditions. This is body modification aimed at specific ends, but facial feminization is weighted differently than other gendered bodily practices like hair styling or makeup application. Candace West and Don H. Zimmerman approach gender as a "routine, methodical, and recurring accomplishment" (1987, 126). In other words, gender is an iterative process, enacted through dress and style, voice and gesture, along with roles and statuses. In short, "doing gender" is the way in which gender comes to be seen as a salient and recognizable social category. To be sure, facial feminization is a way of "doing gender." It is a way that the transwomen effect femininity, but it is unlike other aesthetic means of feminizing. Makeup, for example, is used to give the face a feminine, and thus a more socially valued, appearance. By contrast, FFS is positioned as a mode of repair, not simply a means of "looking better." As a technique for fixing something defined as flawed, in this case disfigured, FFS is understood as necessary. It is not simply gender that gets constructed through facial feminization but rather a bodily stigma that becomes managed.

The stigma, in this case the masculine face, is a product of gendered expectations. To a large degree, a masculine-appearing face can be defined as disfigured for the transwomen because of the stringency of gender. There are two, and only two, socially intelligible gender categories. As Judith Lorber writes, "In Western societies, we see two discrete sexes and two distinguished genders because our society is built on two classes of people, 'women' and 'men.' Once the gender category is given, the attributes of the person are also gendered: Whatever a 'woman' is has to be 'female'; whatever a 'man' is has to be 'male'" (1993, 567).

Transwomen, by the very nature of transitioning, already challenge the social prescription that one is born sexed and that one's gender follows from one's sex assignment at birth. Facial feminization is an enactment of gender—both a means of "doing gender" and a means of thoroughly and resolutely succumbing to a discrete notion of femininity. Because gender is conceptualized as a discrete category, doing gender is compelling, but it is also constraining. In other words, doing gender (and doing it in particular ways) is compulsory, unless one is willing to accept the costs of resisting. In this way, facial feminization and other techniques of transitioning from one gender status are not simply elective in the sense that they are desired. Rather, modes of gendering like facial feminization get chosen partially because gender difference must be contained. Facial feminization works, thus, as a technique of normalizing gender variance.

As opposed to elective or optional interventions, techniques of normalization are directed toward avoiding consequences related to an undesirable or unsustainable social status. Gender can be the grounds upon which normalization proceeds. As West and Zimmerman note, doing gender is required: "Doing gender is unavoidable. It is unavoidable because of the social consequences of sex-category membership: the allocation of power and resources not only in the domestic, economic, and political domains but also in the broad arena of interpersonal relationships" (1987, 145).

In other words, social life depends upon doing gender, and not simply any gender, as West and Zimmerman seem to suggest, but rather socially acceptable modes of gender. Lorber further outlines the consequences related to rejecting or living outside of normative gendered statuses: "Political power, control of scarce resources, and, if necessary, violence uphold the gendered social order in the face of resistance and rebellion" (1993, 578). As Lorber asserts, there are consequences for challenges to the gendered social order.

For transwomen, masculine facial appearance is a direct challenge to sexual dimorphism, to the culturally dominant model upon which gender relies. The narrative that positions the masculine face as a disfigured face maps gender and normalization onto one another. The masculine face is not simply unattractive; rather, it is untenable. Put simply, because gender is a heavily enforced normative category, facial feminization serves as a technique of normalization. What a surgeon does to facial feminization patients is give them faces that make it possible to lead a "normal" life.

Ugly Conclusions

Biomedical technologies and contemporary medical practices are infused by extant notions about gender, and, in fact, science and medicine are increasingly in the business of reinventing hegemonically gendered bodies. In other words, technologies that facilitate transgendering both rely upon stringent conceptualizations of gender and reproduce bodies that reflect radically essentialist notions of sex. For example, without strict ideas about what constitutes a female face, difference in facial appearance cannot be characterized using the language of disfigurement; thus, it is precisely the inadequacy of transwomen's faces that inspires intervention. At the same time, this intervention further entrenches hegemonic bodies. Medical technologies aimed at transsexualizing the body put this practice into sharp relief, but this holds true even for "biological" females and males who are no less engaged in gendered body projects.

Consider cosmetic surgery. Whether attractiveness or normalcy is sought, the practice of identifying and crafting a changed appearance is similarly inspired by the coercive dimensions of gender. It is not surprising that cosmetic interventions become couched in the language of need, as in "I need breast implants." Such a claim only works when the operating assumption is that to be a woman is to have breasts. In short, in many cases in which medicine works as the technological handmaiden to the precarious cultural project of embodying gender, the categories of male and female become more deeply entrenched as static statuses that must be embodied in order to be perceived and to feel normal. When, as in these cases, intervention is so thoroughly premised on deeply static notions of gender, interventions are represented and consumed as necessary rather than elective procedures. Ultimately, then, deploying such logics in the field of medicine is, in effect, coercive, given that patient consumers are not presented with an alternative.

Ironically, medicine itself contains moments wherein hegemonic notions of gender are challenged, and in the process the categories of male and female become opened up. While medical interventions that are aimed at inventing gender rely on stringent notions of sex/gender, medical practices focused on mediating gendered conditions rely on much more fluid notions of who counts as male and female. Consider treatments for endometriosis or prostate cancer, which sometimes employ hysterectomy and prostectomy, respectively. These are interventions that fundamentally intercede in those body parts that constitute maleness and femaleness, and yet the practice of medicine contains within it a logic that opens up what kinds of bodies might be readily perceived as intelligibly gendered.

By way of illustration, consider the popular debate surrounding the case of the world's "first pregnant man." One of the most oft-cited ways of discrediting Thomas Beatie's contested claim was that as a female to male transsexual he has a uterus. Critics insisted that women, not men, have uteruses, and thus Beatie, while appearing manlike, is "actually" a woman (Currah 2008). This claim was not surprising to feminist and queer studies scholars, who are well versed in the essentialism that pervades popular culture. Taken to its logical end though, this argument suggests that women post-hysterectomy have "lost" a fundamentally female trait. Similarly, it reasons that prostate removal could elicit a similar category crisis. Of course, medical practice makes room for such bodily transformations without imposing a radical alteration to one's gender identity and status. Ironically, such examples suggest that there is room for messier notions of sex and gender within the world of science and medicine and that, in fact, routine medical interventions require flexibility around what bodily formations still fall within acceptable notions of sex and gender. The critically important places to continue looking, analyzing, and critiquing are the moments within contemporary medical practice wherein the hegemonic theory of sex difference dominates the rhetorical landscape. In these spaces, difference will always be conceptualized as abnormal, dysfunctional, untenable, and the imperative to repair will prevail.

NOTES

1. I use the word "transsexual" to specifically describe those who pursue medical technologies (surgery, hormones, cosmetic surgery, etc.) to craft gendered bodies.
2. Other surgeons practice techniques akin to facial feminization surgery, but the four I discuss in this chapter treat facial feminization as a unique and intelligible set of surgical interventions.
3. A dremel is a high-speed rotary tool used for grinding, drilling, and sanding.
4. I use the term "transwomen" to refer to male-to-female transsexuals.
5. Observation took place at two conferences. Both feature seminars focused on facial feminization, but each varied relative to the demographics of conference attendees. Southern Comfort, founded in 1990, occurs annually in Atlanta, Georgia. While attendees overwhelmingly identify as transwomen, increasingly the conference attracts transmen and gender-queer attendees. By contrast, the International Foundation for Gender Education (IFGE) Conference, founded in 1986, which changes venue annually, is dominated by transwomen transitioning in midlife.
6. Each of the surgeons I observed is male. This should not necessarily be surprising given that men disproportionately pursue plastic surgery specialties, but it does make for interesting, and sometimes concerning, gendered dynamics between surgeons and transwomen patients. Specifically, male facial feminization surgeons often offer their personal opinions about appearance, and sometimes interactions between surgeon and patients take on flirtatious overtones. In the context of medical care, these patterns should be thoroughly interrogated.
7. The degree to which it makes sense to emphasize the importance of a single pamphlet must be contextualized. First and foremost, Dr. Ousterhout is often attributed with articulating facial feminization methods. His work appears to be the first account published in a peer-reviewed medical journal (1987). Secondly, given the fact that there are so few doctors from which to choose, there are relatively few sources of information through which potential patients can understand what constitutes facial feminization. While I do not have the data to definitely determine the meanings patients attribute to the sources of information, there is evidence to suggest that Ousterhout's pamphlet is widely read and discussed by potential patients, in that it is routinely mentioned on Web sites on which transwomen chronicle their experiences transitioning and discuss possible interventions.
8. The text that appears throughout as data has been excerpted from field notes taken while attending Southern Comfort and IFGE. I use pseudonyms when quoting from my field notes. This attempt at anonymity is a move to emphasize that the story is not about the personalities in this site but rather the narratives that operate.

REFERENCES

Becking, A. G., D. B. Tuinzing, J. J. Hage, and L.J.G. Gooren. 2007. Transgender feminization of the facial skeleton. *Clinics in Plastic Surgery* 34:557–564.

Blum, V. L. 2003. *Flesh wounds: The culture of cosmetic surgery*. Berkeley: University of California Press.

Currah, P. 2008. Expecting bodies: The pregnant man and transgender exclusion from the Employment Non-Discrimination Act. *Women's Studies Quarterly* 36:330–336.

Dull, D., and C. West. 2002. Accounting for cosmetic surgery: The accomplishment of gender. In *Doing gender, doing difference: Inequality, power, and institutional change*, edited by S. Fenstermaker and C. West, 141–168. New York: Routledge.

Etcoff, N. 1999. *Survival of the prettiest: The science of beauty*. New York: Doubleday.

Gould, S. J. 1981. *The mismeasure of man*. New York: W. W. Norton.

Haraway, D. 2004. *The Haraway reader*. New York: Routledge.

Lorber, J. 1993. Believing is seeing: Biology as ideology. *Gender and Society* 7:568–591.

Oudshoorn, N. 1994. *Beyond the natural body: An archaeology of sex hormones*. Thousand Oaks, CA: Sage.

Ousterhout, D. K. 1987. Feminization of the forehead: Contour changing to improve female aesthetics. *Plastic and Reconstructive Surgery* 79:701–711.

Pitts-Taylor, V. 2007. *Surgery junkies: Wellness and pathology in cosmetic culture*. New Brunswick, NJ: Rutgers University Press.

Schiebinger, L. 1986. Skeletons in the closet: The first illustrations of the female skeleton in eighteenth-century anatomy. *Representations* 14:42–82.

Sullivan, D. A. 2004. *Cosmetic surgery: The cutting edge of medicine*. New Brunswick, NJ: Rutgers University Press.

Tavris, C. 1993. *The mismeasure of women*. New York: Simon and Schuster.

West, C., and D. H. Zimmerman. 1987. Doing gender. *Gender and Society* 1:125–151.

11

The Proportions of Fat in Genetics of Obesity Research

SHIRLENE BADGER

If taken separately, the meanings conjured up when the word "gene," or the word "obesity," is headlined hold a unique and powerful currency in the current public imagination. For example, the gene has become an instantly recognized symbol of the recent developments in the discipline of science. The oft-told story of Frances Crick running in to the Cambridge pub the Eagle declaring the discovery of "the secret of life" (Watson 1999) has become a cultural trope of our era. As much as it can be said that the gene is featured as a cultural icon, in similar fashion the current "problem" for public health is the alleged increasing incidence of obesity at a population level. We are declared to be "waging a war" on a disease that has reached "epidemic" proportions, and its consequences feature with almost regular monotony in our print and television media. With obesity, the focus is on behavior: what we eat, where we eat, how often we eat, and similar questions about exercise. It is imbued with moral overtones, stereotypes, and prejudices. Separately, discussions about the gene and obesity could be said to encapsulate two ends of a spectrum, a sort of dichotomous example of the relationship summed up in debates such as those about nature and nurture. To bring the two together potentially marks a rupture in how either has popularly been described. Yet, it is with the junction encapsulated in the bringing together of the concept of the gene and obesity that this chapter is concerned.

In this chapter, I will draw on historical and fieldwork data collected as part of an ethnographic study that crosses the various locales and experiential perspectives of actors in a genetics of obesity study (GOOS). As such, this chapter pivots around GOOS and its research interest in the identification of human obesity disorders resulting from a genetic disruption of the leptin-melanocortin pathways. The 1994 discovery of the role that leptin played in producing weight loss in extremely obese mice was soon after realized by GOOS in extremely rare cases of Mendelian-inherited human obesity. Leptin is believed to operate as a longer-term signal that integrates with other feeding signals and interacts with receptors within the brain to regulate appetite. In order to elucidate and examine the functions of these pathways and their various signals, GOOS follows a clear methodological strategy of recruiting the extreme phenotype. Hence, the younger the child and the more extreme the obesity, the better, they believe, the biological model. Their goal is not merely to find genes implicated in signaling pathways responsible for the control of energy homeostasis, but to further understand the function of those pathways by examining the conditions in which error occurs.

In this chapter I explore both the historical foundations for GOOS, with an emphasis on the model of human development encapsulated in the child body, and recount particular ethnographic moments from my research with them in an attempt to thematically highlight the idea of proportion. By *proportion* I mean the many ways that actors describe, enact, and lay claim to gendered accounts of the relation of parts to a whole. There is an undercurrent in both the personal/familial and reflexive research practice accounts of a struggle for proportion and coherence. Narratives focus on how either the body does not fit "normal" growth patterns, or the research does not fit "normal" research endeavors in this field. In doing so, the claim to "lack of fit" makes explicit the gendered bounds of both science and the ideal body within familial, societal, and, indeed, scientific expectations.

Perhaps I should begin by explaining what I mean in a very pragmatic and contextualizing way with a comparative nod to the work on breast cancer genetics. An important initial point to note is that in the discovery of a genetic mutation that causes obesity (as with other genes such as BRCA I/II), while the genes identified are not sex-linked in terms of inheritance patterns, the condition (like breast cancer) is understood as having gendered implications in social terms. Breast cancer predominantly affects women, and there have been numerous studies on the psychosocial impact of breast cancer on body image. Furthermore, feminists claim that women bear disproportionate responsibility for the health and well-being of society, and sufficient evidence exists to suggest that women are more likely than men to undergo adult genetic testing (Burgess 1999). With breast cancer, this is, in part, indicative of the belief that breast cancer is a disease that only affects women. However, women may also undergo testing to obtain information for their children (Hallowell 1999).

Similarly, for obesity, the importance of the effect of obesity and weight management on body image concerns for women (see, for example, Orbach 1998; Wolf 1991) and the gendered roles in the promotion and adoption of varying dietary practices within the family context (Lupton 1994) have been well documented. This leads us to an interesting side question: To what extent can the genetics of obesity and its history tell us something about the process of *geneticization,* defined as the process when genetic explanations are privileged and people are reduced to their DNA codes (Lippman 1991)? Given that historically it has been widely recognized that there is an inherited component to some breast cancers, and that there is arguably an over-subscription to the belief of what percentage of breast cancer is attributed to a genetic explanation, following the geneticization argument, we would assume that obesity as a high-profile disease would be receiving similar attention in regards to genetic explanations. While medical textbooks are obviously being rewritten to incorporate new genetic explanations of obesity, the current media hype surrounding the "obesity epidemic" holds very little news of genomics. What does this tell us about how obesity has come to be understood as a genetic problem and the balance of input and output in biologically and socially gendered terms?

In order to highlight these themes in more detail, the remainder of this chapter will focus on a succession of rather limited and selective histories that highlight a focus on the specifically processural nature of "normal" and the interrelated, threatening trajectories if proportion is not achieved. These histories are interspersed with particular ethnographic moments and qualitative data that seek to highlight the lived experience of such lack of fit. The first is concerned with the history of endocrinology and normal measures that are the disciplinary foundations for GOOS, and the second focus highlights the history of childhood studies and its inherent forecast of human development.

The History of Endocrinology

Many histories of endocrinology or its substances begin with the notion of mystery (Bankoff 1947; Medvei 1993). They mark the historical and cultural superstitions, practices, and beliefs of other places where key organs were/are believed to represent different virtuous and desirable qualities of character. George Bankoff notes that "even before the dawn of the Christian era, the magic of the human organs figured in folklore and the superstitions of primitive peoples. The tribal warrior knew that he could increase his strength or banish his fear, or face with courage his enemy by eating human or animal organs" (1947, vi). The movement between substance as object and metaphor is obvious. These desirable substances were believed to sustain the fundamental attributes linked with self-preservation and reproduction that were prevalent across all species and, as such, involved the shared observable practices of and beneficial beliefs in eating hearts, livers, brains, and placentas.

But more than marking mysterious qualities, by the end of the nineteenth century these organs were believed to play an important role in the function of equilibrium within the animal body. These qualities pertained to the very essence of life itself. As Victor Medvei notes (1993, 3): "The living organism is a very complex system. To stand up to the dangers of life, to feed, to digest, to procreate, to be born, to survive, to grow and to develop and—all the time—to remain in an equilibrium which is satisfactory, it needs the ability of a quick decision and action."

Within such characteristic excerpts, we begin to see the basic understanding of human physiology as highlighted by endocrinology: the body as a complex interplay between triggers, messages, and rapid responses. The balance of this fragile equilibrium relied on all these processes functioning correctly. Accounting for error was, therefore, a key topic for investigation. From the 1890s, the internal secretions of certain organs became increasingly important in the study of the physical processes of the human body (Oudshoorn 1990a, 6). The eventual conceptualization of hormones—coming from a Greek word meaning "to rush, to set in motion" (Poulakou-Rebalakou and Marketos 2002, 58)—rewrote basic understandings of (ab)normal functioning of the human body. Hormones are thus explained as the chemical messengers that coordinate all fundamental activities of the different parts of the animal body (Baulieu and Kelly 1990). In similar ways to which we have seen a great deal of hype and excitement over more recent discoveries pertaining to DNA, hormones were believed to be the messengers of life, belonging to the "glands of destiny" with the miraculous power to order the whole of our lives (see Bankoff 1947, 18). The well-known current phrase, "it's all in the genes," is arguably a resonating echo of the previous claim: "it's all in the glands."

The nineteenth-century foundations of endocrinology rely not only on the magical possibility of hormones and endocrine glands, but also in the experimental processes informed by a correlation of the functions and responses between human and animal bodies. Physiologists continued the tradition of laboratory practice with the introduction of laboratory animals as important subjects for the investigation of cases of error. The movement from the laboratory to the clinic was characterized by what some writers report as a tradition of experimental endocrinology (see Dale 1963) and a focus on therapeutic outcomes. The treatment of certain human cases through the injection or oral administration of extracts from animal substances became regular experimental practice: whether an extract of sheep's thyroid to treat myxoedema or the filtered juice of guinea pigs' ovaries for uterine affections and hysteria (see Dale 1963; Oudshoorn 1990a). Through these experiments the functions of endocrine glands and the related role of their hormone messengers were believed to be something that could be deciphered and consequently corrected. Bankoff highlights these foundational assumptions in the following excerpt:

> Thus it seems that, underneath this diversity and uniformity, there is the same foundation. For every animal there is a basic pattern of behaviour, and this is true whether the animal be man [*sic*] or field-mouse. It would not be unreasonable to jump to the conclusion that there must be in all

> animals some sort of system or mechanism whereby the reaction to certain situations is controlled and directed. It has taken long and patient research on the part of scientists to obtain the information that enables science to agree with this rather bold assumption. But today it is known that certain characteristics in both animals and men are under the influence of such a system. It is the system known by the name of *endocrine glands.* (1947, 17–18)

The belief in a general body of laws that governed living systems *required* an experimental approach that came to feature within a general medical paradigm at this time. The belief that the body was something that could be known as an objective and coordinated given prefaced the intellectual investment of experimentation. These histories and the significant changes from a biomedical focus on health to normal functioning have been documented by a long tradition of significant social science research. Of particular note is Georges Canguilhem's (1978) examination of the history of the distinction between the normal and the pathological and the move toward a self-conscious, scientific medicine. This was followed by Michel Foucault's (1963) historical investigation of the restructuring and reform of the epistemological field of medicine during the late eighteenth and early nineteenth century. Furthermore, Adele Clarke (1987) and Nelly Oudshoorn (1990b) have explored the more specific histories of sex hormones. Current social science literature (including psychology and psychotherapy) that follows endocrine practice has tended to follow this tradition and focus on the various work and networks of sex hormones. It is worthwhile noting that a sociological history of the work of growth hormones and a detailed social history of the scientific translation from hormones to genes, as is evidenced in the discovery of *ob* or "fat" genes, are obvious gaps in these literatures.

Normal Measures and Extreme Phenotypes

In terms of hereditary predisposition, the histories of obesity, compared to breast cancer or Huntington's disease, are somewhat different. When we turn to obesity, the only historically significant examples of familial pedigrees of excess are in the recorded cases of giants. The Christian Bible records pedigrees of giants, and throughout history there were known genealogies through which extreme giganticism could be traced. However, it is important to note that the cases of giganticism were recorded in specific ways, claiming a significant racial or ethnic structure rather than familial significance, or as an evolutionary link between gods and humans. Historically, giganticism was linked to territory and populations, such as to the Fyrdafjord District of Norway and of South Sweden (Medvei 1993).[1] Importantly, giants were believed to be impotent (and were generally reported as male); thus, the historical moments of giganticism seemed to appear and decline with the rise and fall of those territories and populations.

As Medvei (1993) highlights, an important aspect of endocrinology is distinguishing human variation from disease, with a particular interest in atypical cases. In the case of growth, it seems that the measure of "normal" was what was needed to preface those cases that were atypical. Or the *ab*-normality had to be extreme and self-evident. Hence, measurement of the body was one of the general endeavors practiced, alongside physiology, with the goal of establishing various rules about the nature of the animal and human world. The familiar historical narratives that point to John Lavater's (1855) catalogue of personality traits assigned to physical signs and Cesare Lombroso and Guglielmo Ferrero's (1895) measurements of female deviants are cases in point. However, the point I am trying to make again is that the strategy for investigation pivots around the ideas of peoples, populations, and their norms and deviations. In the case of genetics of obesity, the starting point is individual error within given populations rather than familial and genealogical investigations of illness suffering. To explain further, I will provide a brief history of the development of the indices that retain value for the measurement and categorization of obesity.

Indices of overweight and obesity have, unsurprisingly perhaps, emerged from measures of the laws of averages. Ian Hacking (2007) provides a fascinating exposition of the equations that have come to be known with such banal significance as signaling "*weight*" that they have almost been "black-boxed." In his history of the BMI (body mass index), he begins with the studies of the laws of mankind. Typically, he references Immanuel Kant's *Anthropologie* (1798) before highlighting what he claims to be a similar benchmark in the study of humankind as a species: the law of growth according to Georges-Louis Leclerc Buffon's *Natural History of Man (1744–1788)*, concluding the third volume of his *Histoire Naturelle.* Buffon's law of growth, represented by the equation weight/height^3, was based on the regular measurements of his neighbors' son from birth. It is important to note that these were principally measurements of height, never weight, yet this is regarded as the first study of growth and outlined a pattern that became familiar in the Child Study Movement. Hacking refers to Buffon's resulting equation as Quetelet's artifact. Through the recorded measurements of various male populations, Adolphe Quetelet argued that Buffon had gotten it wrong and proposed the equation weight/height^2 as a better representation of the laws—not just of growth, but of proportion. Quetelet also argued that both biometrical quantities and moral characteristics reflected a similar distribution, as in the now familiar Gauss error curve. Quetelet's interest was not in anomalies, but rather in the laws of averages that could be inferred by vast numbers, and it is these laws of averages that continue to shape the way we understand normal curves today in medical and disease terms. However, the inference of disease from the categorical marks that have been set at various cut-off points in recent history have proven highly controversial. But through all this, the defining focus for laws of growth remains on the laws of averages and their proportions.

In the mid to late 1800s, Quetelet's calculations were extended further by the tabulations of William Banting (who could be classified as the father of the modern diet industry). There is evidence that these tabulations became part of life insurance calculations for risk as early as 1867 in the United States (Czerniewska 2007). But their link to health risks and as an indicator of disease was not to receive popular attention until the period of postwar epidemiology and the wave of studies even more recently that began to study BMI measurements alongside mortality (see Gard and Wright 2005). What is clear is that despite the contention that circulates this measure, the BMI remains a population measurement and should be treated as such. In this sense, the BMI is not a measure of adiposity or obesity; it is a measure of normalcy that signals individual extremes. But the extremes that certain BMI results signal have not been the domain of medicine or disease status until very recently. In fact, throughout the 1960s to the turn of the century, the identification of obesity through BMI scores was immediately passed onto the domain of psychologists and self-help literatures. In this sense, the BMI marked out an extreme disturbance in psychological functioning rather than biology and was not taken seriously by medical professionals or research scientists as biological error.

Reflection

Within the context of research on genetics of obesity, the positioning of fields of research and the legitimacy of scientific claims pertaining to methodology and causation were repeatedly sites of contestation. The justification of interest in individual error over population-based interventions into obesity was a significant task for GOOS. Tales of serendipitous discoveries of error in both mouse models and individual children were the topic for laboratory meetings and were retold over coffee as anchor moments for deciding whether to explore a particular child case further. At these moments the flow of the historical underpinnings of experimental endocrinology featured in simplistic methodological reminders of value. However, the biggest issue remained in proving biological error and the question of how to actively dispel the accusation of other symptoms or character flaws as confounders in obesity causation. Consequently, by the self-evidence of the abnormality, the extreme phenotype is seen to automatically code for error or difference. The ability to reduce to a single biological cause explains the extreme abnormality and removes the possibility of other cultural or social confounding factors. The transformation of a visible condition that serves as a marker for social and moral failure into a single categorical mark on a graph with clear parameters and symptoms is fraught with complexity.

The most banal expression of this throughout my research was how the model of normal human development and a mundane example of a variation on Quetelet's "average man" is captured in the size = age assumptions of children's clothing. Clothes repeatedly served as a signal of weight out of context. These mundane measures of normality serve an interesting ordering function in society. In this research, the talk of time and its passing is frequently measured by children's bodies alongside and in comparison with other children. It is captured in the cry of parents that their child's growth seems boundless and "when will it stop?" It is also captured in the temporal distortion of clothing when produced according to the above-mentioned size = age equation. However, this distortion also served as a marker for something else. As we shall see in the remainder of this section, growth out of size had a particular gendered implication. This was played out in understandings of inheritance, gendered biographies, and symptomology.

Perhaps most striking in this study of a genetic condition was the finding that obesity through the generations was not something that was seen as a problem historically. And, historically within the family, it was especially not seen as a problem in childhood. It was "part of who you are" and part of your (often gendered) biography. Carol, her two sisters, and their four children highlighted this historical indifference to obesity as a "problem" in their retrospective piecing together of a possible familial line of inheritance. Carol highlighted the gendered and generational biography when discussing her family, particularly her mother:

S.B: Was it ever talked about in your family?

CAROL: No, because I don't think anyone ever—but my Mum was about size 18, 16 almost, though if I think because she died so young it was hard. But I'd say she was about size 18, but I can remember she used to eat off of a little tea plate and have her meals off of that, so if I think about it, it must have come from Mum's side because Mum was like that, she hardly ate anything but she wouldn't slim back down, but she'd had eight kids, so them days you used to think, well, you know, you can't expect to have eight kids and not. Do you know what I mean? That's what people used to say: "God you've had eight kids." You know it's like if she had eight kids and went down to a size 10, you'd comment on that. But with Mum she stayed the same. And I can remember Mum's Mum was chubby as well. But it's not until this come out because you never thought of weight. It's not a thing you think of, is it?

Carol clearly marks weight as something that is not distinct as a "thing" that was talked about or thought about problematically in terms of inheritance. Instead, what would have been regarded as problematic after eight pregnancies was weight loss. For Tom and Sally's family, it had always been just a part of who they were. They talked of the "freak" in their family as being their son, who was "like a beanpole," not their daughter, who received a diagnosis of a genetic mutation that caused her obesity. You couldn't expect to have skinny children when you were both big yourself. Weight, like height, eye color, and hair color, was an expected visual representation of belonging in the family. Out of the familial context, weight potentially came to be read as lack of proportion against other biological or social markers. Perhaps, more significantly, weight was also always a personal narrative. In the case of Carol's mother, it was explained by her eight pregnancies and the resulting impact on the female body. Such biographical events were also drawn on to place an emphasis on the sometimes rather complicated beliefs that the gene mutation was carried along female lines. In this sense, fathers were often called "carriers" if they had tested positive. Other interviewees, like Glenda and Marion, told me how in their respective families it had only "come out in the girls." For Tom and Sally's daughter Kimberley, even though her father also had the same diagnosis, she believed the gene to be expressed phenotypically only in the females in the family. This concerned her when it came to her then current pregnancy. She was carrying a boy and told me, "I hope he has got it also, but he probably won't because he's a boy, but so Josie isn't the only one." These familial beliefs about transmission signal that historical and cultural understandings of obesity and expected gendered body shapes and biographies retain dominance over the power of the icon of the gene in today's culture.

Weight problems seem to move fluidly across other medical symptoms in ways that become either visible or invisible explanations for the problem. This fluid quality to obesity enables it to transform into different explanations in different circumstances, as was apparent in the lived experiences of these families. For example, participants repeatedly listed off their medical history like a checklist of all that had been investigated, as the following excerpt illustrates:

WENDY: I've had a lot of medical problems. Like I've got a hernia. I've got gallstones. I did have an overactive thyroid, but that sorted out after I had the twins. And also I was diagnosed as epileptic; yes, the pregnancy triggered off epilepsy. The sickness triggered it off; you know, when you trigger things off in your head; those, um, hormones, it sort of triggered it off. Yeah, the pregnancy triggered it off. They said it's just a chemical imbalance that caused it. It was probably the scans, you know, not scans, the old ECT.

During the interview, Wendy also discussed problems she had experienced as a child, such as psoriasis and dyslexia. However, her focus remained on those problems post childbirth. The inference for the stomach became obvious later in the interview when she claimed:

WENDY: I've got a medical problem with my stomach, so I've got to lose the weight to have the operation. I've got an umbilical hernia. It's my fault for having big babies.

S.B: It's your fault for having big babies?

WENDY: Yeah, I have big babies. Does your stomach in.

She continued the interview, explaining that she had caesarean sections to deliver her four children, and concluded: "It's disturbed down there and it's all come up here [pointing to her stomach]."

In this sense obesity is given meaning in terms of other symptoms. Responsibility for the "medical" problem becomes a complex chicken-and-egg web of what comes first. For Wendy, fat is both a code and a substance. It signals other bodily dysfunction, but as a substance it can also move from the original site of that dysfunction to be exhibited elsewhere in the body. It is both a result of something else, at times an expected outcome of female bodily corporeality (especially if one has children), and a separate problem. Like the self-evident expression of fat, the location of the problem was often visible externally or located in the mouth, the stomach, or the bowel regions. Investigations consequently revolved around measurements, weights, eating behaviors, exercise, and stool samples.

For the girls and women in this study the malleability of the body and, specifically, of fat was regularly tied to pregnancy, gynecological problems, or rare and inconsistent periods and their related investigations of polycystic ovaries, precocious puberty, cysts, and cancer. Fat and blood were only linked metaphorically and literally where fat was believed to block normal flows of blood. Most notable was the connection between blood and fat as substances that were supposed to be mobile and/or transient. More specifically, blood and fat were often linked as signals of maturation and development. Puppy fat was somehow bled out with the onset of menstruation, marking the progression from girlhood to womanhood in significant ways and clearly positioning fat as a childish condition that one was to move on from. The onset of menarche is an important indicator in endocrinology of the normal function of particular growth hormones. In the case of leptin deficiency, for example, a child will not go through puberty unless treated. Furthermore, there are many interpretations of the fragile link between obesity and infertility that currently inform individual clinical policies pertaining to access to intervention such as in-vitro fertilization.

The next excerpts from separate interviews with a mother and her daughter highlight this rather significant relationship between blood and fat in puberty and development:

JOYCE: Then she put another few pounds on and then also she started her periods, so I thought, oh this is it, we're on track! But then nothing [no menstruation] happened again for months and months and months. So the weight was still going on. I say it was compounding, piling it on.

REBECCA: I just want to have another period so I can just bleed it all out. Then it will be fine.

Joyce and her daughter Rebecca, in separate interviews, link Rebecca's irregular menstrual cycle to her exponential weight gain. This reflects a particular view of the body (not dissimilar to endocrinology) where the body is seen as a complex system of flows and messages such that when there is disruption in one there is an ensuing disruption to the flow of another. The onset of menarche is yet another indicator of the normal line of human development that in this case was being somewhat challenged by fat. This is articulated poignantly by Joyce with her exclamation of "we're on track!" But then with no further periods, she sees her daughter's weight continue to pile on. For Rebecca there is a belief that blood and fat can somehow merge and are located in a similar region so that everything would be solved by an anticipated period and thus removed from her body to restore normalcy.

The Duality of Childhood

The inclusion of and focus on children and young people in this research makes reference to the sociology of childhood and obviously acknowledges childhood as a distinct category with different possibilities for meaning and experience. Many writers have examined the historical trends in the social constructions of childhood (Anderson 1980; Ariès 1962; Cunningham 1991, 1995; Hendrick 1997; Heywood 2001; Pinchbeck and Hewitt 1973; Pollock 1983; Shorter 1976; Steedman 1995; Stone 1977; Walvin 1982). Following Philippe Ariès's historical work, there is general consensus that from the Middle Ages, a significant change has occurred that brought into existence specific ideals about modern childhood (Prout 2005, 9). The changes both in beliefs and institutional practices reflect the familiar sociological literatures that describe economic modernization and the changing emotional economy of the family. Throughout the centuries, childhood came to be recognized as something distinct from adulthood. Pia Christensen (1994) has argued that modernity has constructed childhood as the "cultural other" of adulthood, and we see this reflected in a series of dichotomized qualities regarding public and private domains, nature and culture, dependence and independence, play and work.

The dualisms of childhood have long held significance in such domains as education, legislation, pediatric medicine, and psychology. Dominating Western thought has been a model of child development based on the processual character of natural growth that, in turn, connects both biological and social development. Both in scientific and sociological research on/with children the "natural" has arguably been separated from the cultural. The very language of nature evokes dichotomies and is structured on deeply embedded ideas captured in pairs of related opposites (see Jordanova 1986; Williams 1975). Alan Prout has argued that "the history of childhood studies describes a trajectory through this relationship, which, because it has operated within a modernist field of thought that separated culture and nature, zig-zags between the poles of the opposition, now placing childhood at the biological end, now at the social" (2005, 44). In this regard, the child is believed to signify and embody the human model of both natural and cultural facts about life and the process of development.

Many claim that the beginning of modern childhood studies is attributable to the period of the Child Study Movement made famous by the births of William Darwin, Madeleine and Alice Binet, Polly Watson, and Jacqueline Piaget and the observational studies of their development by their scientist fathers (Kessen 1965). Yet others have argued that children have long held a fascination to their respective adults, and Carolyn Steedman (1995) reports that archives are full of the notes and diaries, kept by parents on their children, that are turned into the "science" of child-care manuals. In many ways Charles Darwin's diary on his son William can be seen as one such example and sparked a wave of analyses of single children that came to depict a universability of the nature of childhood.[2] It is no surprise that Darwin's study and those he inspired are framed within implicitly evolutionary frameworks. Yet, as Claudia Castaneda notes, it is not the Darwinian version of evolution that we are most familiar with that is at play here, but rather one that relies on the idea of potential growth. In this sense, Castaneda argues, it is Herbert Spencer's writings about human history evident in the child that dominated public and scientific writings of the nineteenth century. Spencer's philosophy of science "employed a version of evolution that used individual development as the basis for human evolution, and narrated both as a progressive story" (Castaneda 2002, 20–21). The point that Castaneda (following Steedman) is trying to make here is that a particular scientific approach to the physiological body, exemplified by the child body, became the self-evident marker for a range of other developmental, collective, social, and moral claims. Thus, the child then becomes the very representation and embodiment of the problem of growth across the natural world, at one moment savage, in another moment innocent, at one moment animal, in another moment human.

Standing at the borderline of nature and culture, purity and pollution, the child expresses the potential of "Other." The child expresses the relationship between animal and human biologies and explains the characteristics between the savage and the civilized. The potentiality of the child as both category and symbol extends beyond these classifications of what childhood is. Castaneda (2002) and Prout (2005) have argued that these divides have been applied to other concerns of difference such as empire, race, and gender. Despite their inadequacy, the dualisms of childhood have exposed the very mutability and heterogeneity of childhood, indeed, childhoods (see James and Prout 1997, xi). As Ludmilla Jordanova explains, dichotomies have "a marvellous capacity for containing contradictions" (1986, 86). Within clinical research, these dichotomies of childhood hold extraordinary explanatory power. If obesity and the gene mark an important dichotomous juncture, a focus on the child surely confounds these boundaries further.

Reflection

In my research, the pattern of conducting interviews and observing family members from my GOOS sample around Great Britain tended to follow a similar timetable. I would generally meet the mother first and (often) spend much of the day with her. Sometimes I would also interview the father during the day. As the time drew close to mark the end of the school day, I would be invited either to take the car journey to the school gate or to walk the distance with the mother to meet the child under investigation by GOOS for the first time or to watch for them walking home from school with their friends. On the journey I was told various tales about the child at school: the problems and the achievements. I would be prepared for certain other mothers who would also be waiting. Or I would be told about the child's peers with whom I would see them walking. They were notes for comparison in more ways than one: being both a mother and the mother of a peer of the child I was about to meet, or simply being a peer in age and stage. Without fail I would then be told to *look* and *watch* as we waited for the school doors or gates to open: "You will be able to spot her/him straight away. She/he just stands out from the other kids. You won't be able to miss her/him. See? See!" I would watch and wait, in part hoping that I would be able to "recognize" the child I was there for, in part concerned that I wouldn't "see" him or her.

As I watched children run from the school to waiting parents/carers and noted the proportional exception (in this context) of the child I was meeting, I wondered where this epidemic of childhood obesity resided? My interview data is full of accounts of the loneliness of obesity, of being *the* fat kid at school. Loneliness happens within a context that is brimming with meaning and actions from which one is marginalized. To follow obesity, I realized that one must follow it through the different terrains of laboratory, clinical research facility, and home and school life. More importantly, I must follow the child's body through these terrains to understand how and when the body "fitted" and when the body was marked as "different" and out of context. In this sense I follow Mary Douglas's seminal work *Purity and Danger* (1966), where she provides a useful analysis of the rituals of purification and marginality. In order to investigate these rituals we must examine the public categories of which they are a part, for anywhere that there is classification, culture accounts for those people and objects that stand out from its prescribed pattern as an anomaly. Douglas argues that those who occupy a marginal state may not be doing anything morally wrong; however, their status remains indefinable (1966, 96). The margins are seen as dangerous, for dirt and its equivalents challenge defining boundaries and create dissonance between the individual and general interpretations (1966, 40).

The challenge of obesity to both natural and social orders of childhood was highlighted in different ways throughout the study. When I asked many of the children to describe their bodies to me, or to describe what they liked or didn't like about themselves, they would respond by talking about the parts that made them like other family members or they would discursively break their body into parts and their corresponding functions, as the following two excerpts illustrate:

NERALI: My hair is the same color as my dad's and his hair is the same color as my Nan's, and my eyes are definitely the same as my dad's, and hair and just all of me is like my Nana.

HARRIET: I like my hands, they're very useful for feeling and touching and being able to write. I like my nose because I like smelling things, but all year round I have an allergy to pollen. I like my eyes because otherwise you couldn't see where you are. My hair is probably the least important part of my body. It doesn't really do anything.

The younger children in this research liked to talk to me while drawing with the pastels and colored pencils I carried with me. Other than the drawings made during more formal interview moments, they often bestowed gifts of pictures on me as I left their homes or their room in the clinical research facility. While pictures included colorful rainbows and butterflies, the pictures of themselves were curious and were accompanied by a dialogue of quotes like those above of what parts of the body they liked and disliked. Generally, these distinctions were not represented in the complete drawing of themselves. They did not draw themselves as fat or draw certain disliked parts like the belly as extended or bigger than other parts of their body. They drew themselves in proportion. They drew themselves in conventional nuclear families, no bigger or smaller than their siblings. They narrated tales of bullying as they drew, with some expressing a desire to be someone else. Frequently, they drew themselves in proportion but avoiding the gaze of the viewer by looking to the side. In many ways they reveal a self-knowledge of who they are and their place. But they also show an awareness of the othering that the visual representation of their body presents through the gaze of the other. In the stories, for example, that Nerali told and her summing view that "It would just be better for everyone if I wasn't fat," she makes clear that her size is a public concern that impinges on all those around her—whether that be her parents, her sister, her grandparents, those children who choose to be her friend, or the man in the local shop who takes her money when she makes a purchase.

The first time I went to Nerali's house, I met and interviewed her mother, Samantha, while Nerali and her sister were at school. Samantha spent a lot of time at the beginning of the interview describing her daughter to me. She said, "When you see Nerali and when you look at her and you'll think uh-oh because her shape is not natural, you know, it's not a normal shape. If she was twenty you'd assume she was pregnant. You know if you just looked at her. She's got these little short legs, so I have to try and find a school uniform that's big enough for her tummy."

Samantha is concerned by her daughter's lack of normal proportions. Instead of "fitting" into a recognizable twelve-year-old body, she provides a prematurely adult and sexualized description of Nerali looking like a twenty-year-old pregnant woman. As Samantha suggests, there were ways to view these bodies as normal. If Nerali were older and pregnant, she would fit her body. But in the context of being twelve years old, she doesn't. To this end, obesity is not like conditions such as achondroplasia, where the body and growth are not in proportion. There are proportions in obesity of both age and size and their associated expectations.

Discussion

The problem with obesity, in both scientific research and lived experience, is that it cannot easily be reduced to a single case of biological error. This is where the juncture between the gene and obesity is problematic. For example, during my introduction to the field and my first negotiating interactions with GOOS, there was anticipation and excitement in our collective discussions. The clinical researcher, who had almost single handedly established GOOS, compared herself to me on more than one occasion. She claimed that I was like her, that I was charting new territory. She claimed to be the only one "normally" talking about this topic and said her science was the soft science amongst hard sciences because "you can't have controls and I am working with humans rather than mice. . . . I can't invent things." The work she does with rare and extreme children is regarded as the soft science amongst the hard sciences of numbers, large sample sizes, large randomized control trials, and the data extracted from blood and DNA and mice models that are arguably more easily managed within the context of the laboratory. The work she does is also regarded as somewhat peripheral to population-based research on obesity and public health consequences. The disjuncture between metaphors of the gene and obesity are consistently played out in the scientific quest for validity. Furthermore, her clinical work deals with the reality of children's bodies and knowing the very fleshiness of those bodies. However, in the search for biological models of error, where the data from children can be compared to the data from mice, gendered biographies must be removed in an attempt to locate and bracket the disease as extreme and monstrous. Thus, extreme obesity (as individual error) becomes the biological and social threat to the trajectories of proportion. Where children are believed to be the "other" of adulthood and potentially the model for adults, this provides a unique challenge to the "natural order of things": what sort of man or woman will this child become?

Obesity in its extremity can sometimes be invisible, for how do I explain to the reader that a two-year-old I met did not look *ab*-normal, but rather looked like an overweight eight-year-old? How do parents explain the behavior of that two-year-old and pushing her in a pram when she looks like an eight-year-old? Where the child body measures time and the passage along the line of human development, disruptions to this are seen as monstrous and threaten the categorizing boundaries of the social and biological worlds. On the other hand, the self-evidence of biological error and, hence, the visibility of obesity as being "out-of-proportion" become necessary for the diagnosis and work of GOOS. For children to be in proportion, their bodies must fit gendered trajectories of the normal line of human development as exemplified in population measures such as the BMI or through the onset of such stages as menarche for girls. In the everyday lives of participants, the claim to a social meaning of symptomology and the search for a biographical moment where proportion could be reached were important for finding balance for gendered biographies. In this sense, the public categories of child as being and becoming are confounded by the visible expression of obesity and the ways in which fat can be normalized (within different gendered explanations) or presented as an anomaly.

The empirical examples of disjuncture expressed between visibility and invisibility, the social and biological, being and becoming, and "soft" and "hard" sciences in this research reveal just how important it is to explore the materialities of diseases, their fleshiness and physicality, alongside the material cultures of the fields of science and medicine that inform them. In this sense, progress along the normal line of human development, growth charts, and medical models of error become just as potent as issues of clothing, pregnancy, and menarche not only for understanding the multiplicity of a condition like obesity, but also for understanding the identities and biographies we narrate in order to seek proportion (see Mol 2002). The relevance of this exploratory study and the simple moments highlighted open up further diverse theoretical debates surrounding the well-known discussions of power and what narrations of human history get told. The stories told here highlight questions not only of causation and the impact of illness, but also of what does it mean to be human and at what stage does illness signal a degeneracy of a particular gendered identity? In this study, descriptions of the obese body hold a significant scientific and personal power. In the scientific story, the disease story, and the personal stories of obesity, the wobbly line of legitimacy is confounded by gendered biologies and biographies and, thus, challenges the ordered boundaries of science, medicine, biology, and culture. The reliance on gendered biographies in the telling of obesity by participants, perhaps, tells us more about who bears responsibility for collating family health information, for remembering and integrating the past in order to build a cohesive narrative. Obesity is not just about a diagnosis, weight, or a BMI score, it is also about the juncture of individual fit in a world of population measures that are based on ideas and histories of gendered proportions.

NOTES

1. The focus on territory and population resembles the methodology of much scientific/genetic research on obesity, where the value of population-wide genome scanning and the thrifty genotype hypothesis predominate explanations. This, I argue, is a distinctly different strategy from the investigation of other genetic conditions that have a starting point of the biological family unit.
2. We have seen this potential even more forthrightly earlier in this chapter, where Hacking (2007) argues that the foundation of the BMI equation lies in a similar study of a single child and the detailed record of his growth (read height). Hence, the study of a child came to inform measurements of the "average man."

REFERENCES

Anderson, M. 1980. *Approaches to the history of the Western family: 1500–1914*. London: Macmillan.

Ariès, P. 1962. *Centuries of childhood: A social history of family life*. New York: Vintage Books.

Bankoff, G. 1947. *The conquest of the unknown: The story of the endocrine glands*. London: MacDonald.

Baulieu, E. E., and P. A. Kelly. 1990. *Hormones: From molecules to disease*. New York: Hermann Publishers.

Buffon, G. 1744–1788. *Natural history of man, the globe, and of quadrupeds*. New York: Leavitt and Allen.

Burgess, M. 1999. Marketing and fear-mongering: Is it time for commercialised genetic testing? In *The commercialisation of genetic research: Ethical, legal, and policy issues*, edited by T. Caulfield and B. Williams-Jones, 181–194. New York: Kluwer Academic.

Canguilhem, G. 1978. *On the normal and the pathological*. New York: Zone Books.

Castaneda, C. 2002. *Figurations: Child, bodies, worlds*. Durham, NC: Duke University Press.

Christensen, P. 1994. Children as the cultural other. *KEA: Zeischrift fur Kulturwissenschaften, TEMA: Kinderwelten* 6:1–16.

Clarke, A. E. 1987. Research materials and reproductive science in the United States, 1910–1940. In *Physiology in the American context, 1850–1940*, edited by G. L. Geison, 323–350. New York: American Physiological Society.

Cunningham, H. 1991. *The children of the poor: Representations of childhood since the seventeenth century*. Basil Blackwell: Oxford.

———. 1995. *Children and childhood in Western society since 1500*. London: Longman.

Czerniewska, A. 2007. From average to ideal: The evolution of the height and weight table in the United States, 1836–1943. *Social Science History* 31 (2): 273–296.

Dale, H.H.L. 1963. Some endocrinological memories. In *Perspectives in Biology*, edited by C. F. Cori, V. G. Foglia, L. F. Leloir, and S. Ochoa, 19–23. London: Elsevier.

Douglas, M. 1966. *Purity and danger: An analysis of concepts of pollution and taboo*. London: Routledge.

Foucault, M. 1963. *The birth of the clinic*. London: Routledge.

Gard, M., and J. Wright. 2005. *The obesity epidemic: Science, morality, and ideology*. London: Routledge.

Hacking, I. 2007. Where did the BMI come from? Paper presented at Bodies of Evidence: Fat across Disciplines Conference, Cambridge, UK.

Hallowell, N. 1999. Doing the right thing: Genetic risk and responsibility. *Sociology of Health and Illness* 21 (5): 597–621.

Hendrick, H. 1997. Constructions and reconstructions of British childhood: An interpretative survey, 1800 to the present. In *Constructing and reconstructing childhood*, edited by A. James and A. Prout, 34–62. London: Falmer Press.

Heywood, C. 2001. *A history of childhood*. Cambridge: Polity Press.

James, A., and A. Prout. 1997. *Constructing and reconstructing childhood: Contemporary issues in the sociological study of childhood*. London: Falmer Press.

Jordanova, L. 1986. Naturalizing the family: Literature and the bio-medical sciences in the late eighteenth century. In *Languages of nature: Critical essays on science and literature*, edited by L. Jordanova, 86–116. London: Free Association Books.

Kant, I. 1798. *Anthropology from a pragmatic point of view*. Cambridge: Cambridge University Press.

Kessen, W. 1965. *The child*. New York: John Wiley & Sons.

Lavater, J. 1855. *Essays on physiognomy: Designed to promote knowledge and harmony among mankind*. London: William Tegg.

Lippman, A. 1991. Prenatal genetic testing and screening: Constructing needs and reinforcing inequities. *American Journal of Law and Medicine* 17 (1–2): 15–50.

Lombroso, C., and W. Ferrero. 1895. *The female offender*. London: T. Fisher Unwin.

Lupton, D. 1994. Food, memory, and meaning: The symbolic and social nature of food events. *Sociological Review* 42:664–685.

Medvei, V. C. 1993. *The history of clinical endocrinology*. London: Taylor & Francis.

Mol, A. 2002. *The body multiple: Ontology in medical practice*. Durham, NC: Duke University Press.

Orbach, S. 1998. *Fat is a feminist issue*. London: Arrow.

Oudshoorn, N. 1990a. On the making of sex hormones: Research materials and the production of knowledge. *Social Studies of Science* 20 (1): 5–33.

———. 1990b. Endocrinologists and the conceptualization of sex, 1920–1940. *Journal of the History of Biology* 23 (2): 163–86.

Pinchbeck, I., and M. Hewitt. 1973. *Children in English society*. London: Routledge.

Pollock, L. 1983. *Forgotten children*. Cambridge: Cambridge University Press.

Poulakou-Rebalakou, E., and S. G. Marketos. 2002. Endocrine terminology in Corpus Hippocraticum. *Hormones: International Journal of Endocrinology and Metabolism* 1 (1): 57–58.

Prout, A. 2005. *The future of childhood: Towards the interdisciplinary study of childhood*. London: Routledge Falmer.

Shorter, E. 1976. *The making of the modern family*. London: Colins.

Steedman, C. 1995. *Strange dislocations: Childhood and the idea of human interiority, 1780–1930*. London: Virago Press.

Stone, L. 1977. *The family, sex, and marriage in England, 1500–1800*. London: Weidenfield and Nicolson.

Walvin, J. 1982. *A child's world: A social history of English childhood, 1800–1914*. Harmondsworth: Penguin.

Watson, J. D. 1999. *The double helix: A personal account of the discovery of the structure of DNA*. London: Penguin.

Williams, R. 1975. *The country and the city*. London: Paladin Books.

Wolf, N. 1991. *The beauty myth: How images of beauty are used against women*. New York: Vintage.

12

Making Male Sexuality

Hybrid Medical Knowledge and Erectile Dysfunction in Mexico

EMILY WENTZELL

Urology patients at the Instituto Mexicano del Seguro Social (IMSS) hospital in Cuernavaca, Mexico, often say that Mexican men are "sex obsessed." These older, mostly working-class men suffered from problems like prostate enlargement or cancer, heart disease and type-2 diabetes, which often hindered erectile function. The IMSS environment, where Pfizer-funded[1] wall posters labeled less-than-ideal erections as the medical pathology "erectile dysfunction" (ED) and the doctors frequently breakfasted with the friendly Cialis sales representative, might seem conducive to the application of medical treatments to sex-obsessed patients' faltering erections. However, IMSS urologists rarely recommended medical ED treatments, and the vast majority of patients considered drugs like Viagra to be "silly," at best, and deadly, at worst. Those few patients who did try ED drugs did not accept the wall chart's suggestion that their erectile difficulty was a straightforwardly biological problem; instead, they usually saw it as a symptom of a complex set of life and health difficulties that kept them from being the kind of man they wanted to be.

This chapter, based on interviews with over 250 male urology patients and ethnographic observation conducted at the IMSS over ten months in 2007 and 2008, shows how applied medical knowledge about decreased erectile function came to look very different from doctors', patients', and ED drug marketers' abstract descriptions of "male sexuality" and "Mexican masculinity." These actors all described Mexican manhood and men's sexual needs as stemming from an essential nature, though different discourses offered conflicting visions of what this nature entailed. Yet when patients and doctors discussed and treated decreased erectile function, they tempered these universalizing discourses with local ideas of aging, bodily health, and the socially appropriate male life course. Drawing from a range of extant discourses about men, sex, and health, they reconciled different notions of the nature of male sexuality and

Mexican masculinity into explanations of patients' specific social struggles and bodily experiences. Their production of applied medical knowledge was also mediated by the ways that structural forces, like class and lack of resources in the IMSS, shaped and constrained these medical encounters. Thus, treatment decisions relating to decreased erectile function in the IMSS revealed that despite constant (if conflicting) assertions from patients, doctors, and drug companies that male sexuality and the essence of the Mexican man are obvious and natural, in practice, patients and doctors work from hybrid and patient-specific understandings of sexuality and manhood.

Anthropologists have long argued that medical practices both reflect and shape the ways that patients understand themselves and their bodies. Ideas of "health" carry with them implicit notions about the ideal way to be a person in society; using medical treatments entails bringing one's body in line with these norms. The ideals enfolded in understandings of health are often particularly apparent in terms of gender, as medical interventions ranging from fertility treatments to baldness cures help people to embody ideal ways of being a man or a woman (Becker and Nachtigall 1992; Szymczak and Conrad 2006). As Western biomedicine has become the dominant healing system worldwide, the process of "medicalization," in which medical ways of thinking are applied to aspects of life previously understood in other ways, has become inherently globalizing (Clarke et al. 2003). When medical treatments are developed in one cultural context but sold worldwide, their application requires patients and doctors to grapple with the notions of health encoded in the treatment and its marketing, reconciling them with potentially disjunctive local ideas about bodies and illness (Appadurai 1996; Nichter and Vuckovic 1994; Ong and Collier 2006; Petryna and Kleinman 2006; van der Geest and Whyte 1989).

Thus, in the case of ED treatment, doctors and patients must reconcile the norms of masculinity and health encoded in ED drugs, those promoted by the drugs' marketing, and local notions about the nature of manhood, sexuality, and bodily function. In generating medical knowledge and practice specific to a particular man's decreasing erectile function, IMSS patients and doctors draw from conflicting discourses about the "nature" of sexual health and Mexican masculinity. This chapter uses the concept of "hybridity," the notion that cultures and selves are constructed through relationships with a variety of different influences, for understanding this process. This approach has long been popular in studies of Latin America, first as a way to understand histories of the cultural and ethnic mixing resulting from colonization (Hewitt de Alcántara 1984) and more recently to understand the influences of globalization (García Canclini 1995) and understandings of illness that combine biomedical and folk beliefs (Daniulaityte 2004; Mercado-Martinez and Ramos-Herrera 2002). Scholars of gender and sexuality in Mexico and its diaspora have used hybridity theory to understand the practices by which people enact gender and make sexual meanings, arguing that instead of choosing one discourse, such as religion, medicine, or "traditional" wisdom, to understand sexuality, individuals draw from multiple discourses in ways unique to their own situation (Amuchástegui Herrera 1998; González-López 2005; Hirsch 2003).

This chapter will show how male IMSS patients and doctors weave different discourses about the nature of male sexuality and sexual health into hybrid forms of applied medical knowledge that explain erectile function change in social context. Influenced but not controlled by competing, universalizing narratives of male sexuality and health, this knowledge makes sense in cultural context but is unique to each individual. This chapter will track the ways that people draw from, combine, or challenge these discourses in their characterizations of masculinity and sexual health, and the ways that various social factors, including the structural forces that shape medical encounters, influence the hybrid medical understandings that result. It will first present two discourses on male sexuality that seem potentially hegemonic in the IMSS context: the notion of "male sexuality" encoded in ED drugs and the idea of Mexican "machismo." Next, it will describe the local discourses of "mature" masculinity and "natural" male aging that conflict with these ideas. After examining the ways that structural forces shape the medical encounters in which doctors and patients apply aspects of these discourses to the bodily changes at hand, this chapter will show how different sets of circumstances encourage the production of particular types of hybrid medical knowledge. It will conclude with a case study of IMSS ED treatment that illustrates one man's efforts to create a hybrid way of understanding his decreased erectile function in the context of his efforts to be a particular kind of man.

Norms of Male Sexuality: ED and Machismo

The concepts of ED and machismo both naturalize forms of male sexuality dominated by the lifelong practice of penetrative sex, although they entail different ideas about the relationships between bodies, masculinity, and culture. ED is defined medically as "the persistent inability to achieve or maintain an erection sufficient for satisfactory sexual performance" (Lizza and Rosen 1999). As a biomedical pathology publicized by Western drug companies, this characterization of decreased erectile function implies that male bodies are fundamentally similar, regardless of age or social context, and that decreased erectile function is a biological problem that should be treated with a pharmaceutical cure. Critics have argued that while this "medicalization of impotence" offers men a way to achieve firmer erections, it has also created new social pitfalls by implicitly promoting social norms of health, sexuality, and masculinity as biological facts (Tiefer 1994).

For instance, this definition of ED promotes phallocentric ideas of sex, implying that "satisfactory" sex requires a hard penis (Loe 2004; Potts et al.

2004). It also suggests that universal sexual norms exist between men and across cultures, and, in tandem with medical treatments for ED, it fuels the medicalization of sexuality by casting difference in sexual function as pathology (Tiefer 1995, 1996). Medical understandings of sexuality impose norms of sexual function, pathologize sexual difference, and can obscure the reality that sexual experience involves social as well as physical factors (Bass 2001; Mamo and Fishman 2001). Further, the idea that healthy masculinity requires unflagging erectile capacity entails the medical assertion that men "naturally" require penetrative sex to be truly manly (Baglia 2005; Loe 2004; Rubin 2004). These norms also cast bodily changes associated with aging itself as pathological, asserting that "health" requires never-ending performance of youthful sexuality, even if costly pharmaceuticals are required to sustain it (Katz and Marshall 2002; Marshall 2007). Thus, ED treatments encode a particular understanding of male sexuality in which ever-present, unfailingly firm erections and the desire to use them in penetrative sex are "natural," and deviation from this norm represents a medical pathology.

ED drugs and their underlying logic have been well known in Mexico for over a decade. Introduced in 1998, Viagra quickly became readily available without a prescription at pharmacies throughout the nation. ED drugs Cialis and Levitra are also widely available, and a generic ED pharmaceutical has been included in the list of drugs that government hospitals must provide cost-free to eligible patients. ED medications have even been dispensed free of charge to older men in Mexico City in a government attempt to raise morale among the aged (CNN 2008). ED pharmaceuticals, as well as herbal copycats like Powersex and Himcaps, are advertised on brightly colored signs posted on pharmacy walls. M-force, the most heavily marketed ED supplement, advertises frequently on Mexican network television. The label Viagra is frequently attached to food items thought to have reinvigorating properties, like the ostensibly aphrodisiac sea urchin "Viagra" soup. ED and Viagra jokes are common fare in television comedies and joking among friends, often in contexts that lampoon Mexican men's supposed obsession with penetrative sex. Thus, men experiencing decreasing erectile function in Mexico cannot help but know that their condition could be labeled as the medical pathology ED and treated with drugs; a recent study found that nearly all of the male IMSS patients surveyed were familiar with ED drugs (Wentzell and Salmerón 2009).

Like the notion of ED, the concept of Mexican machismo naturalizes the practice of penetrative sex as an essential part of masculinity. However, while the biomedical concept of ED casts erection as a biological function that can be achieved by all healthy men, understandings of machismo cast masculinity as an achieved status, which, while shaped by essential Mexican ethnicity, must be proved and demonstrated through dominant acts like sexual penetration. Machismo is a patriarchal style of masculinity marked by high sexual desire, a supposedly innate tendency to womanize, and dominion over one's emotions. Coined in the 1930s, this concept was popularized by poet Octavio Paz's 1950 essays defining the role that conquest and racial mixing play in the "Mexican national character" (Ramirez 2009). Paz's now widespread argument states that machismo was founded at the same moment as Mexicanness itself, as a response to the rape of the indigenous interpreter Malinche by the conquistadors. This coercive sex between conquistadors and indigenous women both spawned the Mexican people and burdened them with a sort of original sin; Mexicans are thus both constitutionally similar to conquistadors and forever symbolically "fucked" by them. Paz argued that the historical shame associated with this founding moment leads Mexican men to seek self-validation through aggressive sexual penetration and to believe that they must never open up, emotionally or physically, since being penetrated is the trait that marks women as oppressed (Paz [1961] 1985). Whether they see it as true or false, Paz's understanding of Mexican masculinity has become a key reference point through which Mexican men understand themselves as men (Gutmann 1996).

Study participants had complicated relationships to the notion of machismo. Many decried it as an inaccurate and unflattering stereotype that fuels racism abroad. For instance, a fifty-year-old truck driver said, "Machismo is a reputation, but it isn't true; we're different. It's bad; others have a bad impression of us." However, the majority of study participants also had surprisingly damning things to say about "Mexican men" in the abstract. Despite the fact that they themselves were Mexican men, they claimed that this group was constitutionally predisposed to machismo and the negative social traits that went with it. For example, a sixty-seven-year-old retired Mexican history teacher both railed against his country's "bad reputation" abroad and said, "Mexico wasn't colonized like the U.S., but conquered. Spain opened jails, sent thieves, criminals over. All men, so they took advantage of the indigenous women. That's why Mexicans are lazy, criminals, thieves, marijuana smokers; because our culture isn't clean. It's mixed." When asked if this could be completely true in light of the fact that he himself was a faithful husband, honest man, and hard worker, he replied, "Of course!"

Participants frequently brought up the notion of machismo as they made sweeping statements about the nature of "the Mexican man." For instance, a fifty-nine-year-old retired bus driver said, "That's how we Mexicans are, a race that—for women, it's something special with women. We want many women, though we don't care for the kids." Similarly, a thirty-year-old government office worker said that he could not be faithful to his wife: "Lamentably, the Mexican man doesn't have this characteristic." Likewise, a sixty-six-year-old retired driver explained the notion of Mexican men's emotional closure succinctly, saying, "I prefer to keep my feelings to myself; that's how we Mexican men are." The notion of machismo and these totalizing statements about Mexican masculinity naturalize a type of male sexuality that is penetration-oriented, divorced from emotion, and uses sex as a key marker of successful masculinity.

Thus, while the discourses of machismo and ED both cast penetrative sex as a key element of manhood, they entail competing views of the nature of Mexican men. In a social context where both discourses are key reference points for men's understandings of their sexual function, different actors juxtapose these discourses in different ways to make specific claims about the nature of Mexican masculinity. For example, these discourses can be brought together to support the claim that lifelong, penetrative sex is both natural for Mexican men and key to Mexican masculinity. Television ads for M-Force, an herbal ED drug knock-off, feature a ruggedly handsome actor explaining, "It's not for the man who can't—it's for the man who wants more." This statement reassures customers that using erection aids will not reveal a weakness in, but instead will enhance, their masculinity; this concern owes much to macho fears of revealing sexual inadequacy, expressed through the ED drug-related notion that one can increase one's potency with pills.

However, actors who adhere closely to one discourse or the other often seek to highlight the distinctions between these two logics. Some participants who understood their sexual and health choices to be motivated partly by machismo said that ED treatment entailed admitting weakness that would compromise their manliness. A fifty-nine-year-old Spanish teacher said, "It's *machista* to think so, but I don't need foreign substances [for sexual function]." Conversely, the marketers of some ED drugs cast their pharmaceutical treatments as modern, scientific health aids that would support companionate sexual relationships and aid in the national fight against *machista* attitudes. For example, a urologist hired by the Cialis representative to give talks that promoted ED treatment framed machismo as a negative, socially backward characteristic. He argued that *failure* to use ED drugs was macho as it reflected an inability to admit vulnerability and would harm men's marital relationships. Contrary to the M-Force sales strategy, ED pharmaceutical marketing often sought to distance its product from the discourse of machismo by naturalizing ED as a medical disease rather than a social problem.

Conflicting Discourses: "Mature" Masculinity and Local Ideas of Bodies and Health

Despite the fact that many study participants used machismo to explain aspects of their masculinity, they also argued that it was not a valid way to be a man in modern society. Most participants said that being macho was undesirable and unacceptable since times were changing, women have equal rights, and men should enjoy emotional closeness with their families. Many older men who extolled the virtues of egalitarian relationships said that they themselves had been macho as youths and were now "ex-machistas." They saw embodiment of machismo as an immature phase to be enjoyed but outgrown. Thus, they proudly described maturing and focusing increasingly on emotional relationships with family, especially as age and illness made it difficult (and unbecoming) to carouse and seek extramarital sex.

For this reason, study participants generally saw cessation of previously enjoyed and valorized practices like promiscuous sex and partying not as a sad consequence of aging, but as a proactive way of adding elements of maturity into one's way of being a man. For many, embodying age-appropriate manliness became a point of masculine pride. In response to a question about whether he continued to engage in sex, a sixty-eight-year-old barber laughed and stated, "Here in Mexico, we have a saying: 'After old age, smallpox.' It means that some things become silly when one is older." Similarly, linking ideas about the importance of sex to manliness and to his youthful performances of masculinity, a sixty-four-year-old retired university staff member remarked, "Now I don't have sex. I don't have the desire. I don't feel that. I don't even try. It gets erased. I don't feel bad—sometimes when I was young, yes, if you couldn't get it, you felt bad, but not now. A youth looks for it, now no." This discourse of "mature" masculinity tempers the universalizing claims that lifelong penetrative sex is "healthy" and "natural," which are associated with machismo and ED. This notion of mature masculinity instead naturalizes cessation of sexual function in older age. In this view, male sexuality is a diminishing drive or finite life phase, rather than a biological imperative that should be acted out even if medication is necessary to achieve erection.

Local understandings of the nature of bodies and health also conflict with the notion of the desirability of lifelong penetrative sex embedded in the concept of ED drug treatment. Since study participants saw the sexual changes that come with bodily aging as natural and appropriate, medicating these changes away seemed like a dangerous distortion of the "natural" life cycle. The older IMSS patients overwhelmingly rejected ED drug use, dismissing the notion that decreasing erectile function in older men was a medical pathology. Based on this logic, participants frequently voiced not only distaste for, but fear of ED pharmaceutical treatment, arguing that treatments like Viagra would unnaturally "accelerate" one's body. Participants tended to see "fast" living as appropriate for young men and youthful sexuality, but dangerous for older, slowing bodies. For instance, a seventy-eight-year-old food vendor stated that ED drugs "accelerate you, to your death. Many friends have told me: they will accelerate you a lot, then you'll collapse; that stuff will kill you." For these men, the inappropriateness of ED treatment during mature masculinity made it physically risky, since it required pushing one's body to perform acts that were no longer appropriate to its place in the life course. Thus, this view of male nature casts lifelong penetrative sex as not only inappropriate, but so unnatural—and thus unhealthy—as to be potentially deadly.

Structural Influences on the Construction of Hybrid Medical Knowledge

As they discuss, diagnose, and consider treatment for urologic problems that involve decreased erectile function, IMSS doctors and patients draw from all these ways of understanding Mexican manliness and male sexuality to construct hybrid, context-specific medical knowledge. Despite the biomedical nature of doctor-patient encounters, the competing discourses described above also profoundly influence the ways these actors understand decreased erectile function, since they bring cultural ideas of gender and health into treatment encounters. In addition to the social salience of local understandings of gender and health, structural factors that shape medical interactions influence the ways doctors and patients weave different ways of understanding decreased erectile function into patient-specific medical knowledge.

Doctors' and patients' desire to temper medical understandings of erectile function change with other culturally salient explanations is particularly strong in the context of the IMSS hospital. IMSS healthcare is a nationwide program available to all formally employed workers in the private sector (INEGI 2005). While comprehensive, the IMSS system is plagued by a lack of resources and a confusing bureaucracy (Carlos 2009). Patients face long waits and logistical difficulties, which encourage people with the means to afford private care to go outside the IMSS for health services. Doctors' need to see a high volume of patients results in rushed and often impersonal interactions and in treatment focused on only patients' most serious ailments. While IMSS patients may not desire treatment for decreased erectile function because they do not see it as a medical concern, the minority that does want this treatment may not be able to obtain it. The lack of doctor-patient familiarity and trust encouraged by the IMSS system often led patients to keep quiet about non-life-threatening health concerns, and may encourage both doctors and patients to favor social rather than medical explanations for bodily changes.

For example, structural disincentives to see erectile function change as a medical problem in the IMSS context encourage the idea that pharmaceuticals promoting youthful forms of embodiment are dangerously inappropriate for older men's bodies. When they discussed ED treatment, IMSS urologists were much more likely to suggest, and patients were more likely to accept, a vitamin than an ED drug. From many patients' point of view, this was a physically safer option that would revitalize their body without the potentially dangerous side effects of ED medications. For example, a seventy-eight-year-old retiree stated that vitamins could work like, but more safely than, Viagra by "strengthening the blood" and promoting "a fertile life," just like the revitalizing vegetable soup he drank at the market. The urologists preferred vitamins both because they were a cost-effective treatment that would not necessarily require medical follow-up, and because they might address the social and emotional factors underlying erectile difficulty. Combining medical and social approaches to the symptom of erectile difficulty, the doctors often said that erectile difficulty might be caused by a "vicious cycle," in which a one-time problem engendered a self-fulfilling fear that the difficulty would continue. The doctors saw the vitamins as physically beneficial, but also believed that they might provide a psychological boost that could help the patient break this cycle.

Conversely, doctors in private settings may encourage medical understandings of non-life-threatening ailments, including erection problems, because they have the time, inclination, and financial motivation to classify them as such. While medical interactions in the IMSS were brief and focused on a few key problems reported by the patient or referring physician, private doctors reported asking their patients questions about their health and lives in general, seeking to elicit a broader picture of their patient's health and pick up on unreported issues like erectile difficulty. However, while private medical care did make it more likely that patients would be diagnosed with ED or offered ED medication, this treatment option did not eclipse hybrid approaches that cast erectile difficulty as complexly biosocial and multifactoral.

In fact, doctors engaging in privately funded interactions with patients used the extra resource of time to delve more fully into their patients' social lives. This often led to doctors characterizing patients' sexual difficulties as life problems and casting ED drugs as a treatment for a symptom rather than the patients' underlying social problems. For example, a private-practice family medicine doctor working in a lower-middle- and working-class neighborhood often prescribed ED drugs to patients that were unhappy with their erectile function, sex lives, or marriages, but cast the drugs as aids that would support behavioral changes, rather than as cures for a strictly biomedical problem. She stated that often her patients live in one room with their entire family—she would recommend that they send their family out to the movies so that they could have private time to engage in sex with their partner. She also recommended that they try non-penetrative acts, like oral sex or manual stimulation. She reported that her older and generally quite conservative patients were shocked to receive this advice, but often tried it, and sometimes returned to happily tell her about their progress. The IMSS urologists also reported combining drug prescriptions with advice aimed at treating the social factors underlying sexual difficulty, especially when they saw private patients outside the IMSS and had more time for conversation. For example, one doctor would advise men to clear their minds of other problems and fully concentrate on sex, while another would recommend that men watch pornography if they were not feeling aroused by their wife.

Of course, the situation would likely be different with patients who sought and could afford treatment from doctors specializing in ED. Clinics specifically devoted to this issue generally cast the problem as straightforwardly medical and provided pharmaceutical solutions. Thus, in medical interactions shaped by different structural pressures and involving men working from different life experiences and physical bodies, patients and doctors drew from a set of culturally intelligible ways of understanding Mexican masculinity and male sexuality to create very different ways of understanding and intervening in changing erectile function.

Hybrid Understandings of Health and Male Sexuality

Differences in these patient-specific forms of medical knowledge stemmed from variation in patients' health and life experience. Even in the IMSS setting, where decreased erectile function was usually not interpreted as a medical concern, study participants who felt that they were successfully "being men" in most areas of life, but had chronic diseases that had clearly affected their erectile function, often adopted largely medical understandings of less-than-ideal erections. A sixty-three-year-old retired salesman saw ED treatment as a remedy for a problem caused by prostate cancer, over which he had little control: "I know that [the ED] is a product of the prostate; it's one medicine more, to keep functioning. That will normalize me." He felt that ED drugs would thus help him return to "the normal—return to my own rhythm of relations." Even when people focused most heavily on medical discourse, they tended to incorporate aspects of locally extant discourses on male sexuality. For instance, a couple who had been using Cialis reported that redressing the husband's physical changes with medicine had allowed them to return to a "normal life." Explaining their adoption of medical norms partly through the local discourse that aging bodies slow and require different care, the wife reported that using an ED drug seems "normal; if there is hypertension, diabetes, it's normal that there isn't erection. When you're young, healthy . . . things change. One has to adapt to one's organism."

However, the majority of men seeking ED treatment aligned themselves more closely with understandings of Mexican masculinity that cast ongoing penetrative sexuality as natural and its lack as abnormally unmanly. They thus understood erectile difficulty as the culmination of multiple failings in their performances of masculinity. Despite their adherence to certain elements of the macho and medical ED discourses, they saw ED drug treatments as a solution to only one aspect of a composite problem with physical, social, and structural causes. Men identified illness and aging, economic and work problems, stress at work and in the home, and faltering social and romantic relationships as causes (as well as sometime consequences) of their erectile difficulty. They did not see themselves to be suffering from a discrete physical pathology, but instead saw ED as an embodied marker of their faltering masculinity.

For example, many participants, particularly those whose romantic relationships had been disrupted by their sexual problems, described the drugs as a psychological aid. They often characterized ED medications as restoring function that had been "blocked" by social and emotional circumstances. A sixty-year-old office worker reported, "The pills were to unblock me." A fifty-three-year-old participant summed up the idea that ED treatment would aid confidence and improve perception of one's own masculine selfhood. He was experiencing erectile problems while going through a separation from his wife, caused by her discovery of his infidelity; he said that what he hoped to gain from treatment was, "more than anything else, confidence. Dysfunction is a problem of the head, of not having confidence." Drawing on hybrids of medical understandings of ED and notions of ideal masculinity allied with the concept of machismo, many men using medical treatment saw it not as a cure for a straightforwardly biomedical pathology, but as an intervention into a bodily symptom of a complex life problem.

José: Hybridizing Medical and Mexican Masculinity Discourses

José's case shows how one man used ED treatment as a forum for combining multiple discourses about health and masculinity in an attempt to understand and mitigate his perceived failure at being a man.[2] A divorced forty-year-old construction worker, José was one of the few men who came to the IMSS urology department specifically for ED treatment. He was very soft spoken, though open about his sexual and other problems, and hungry for treatment information. He understood his erectile difficulties to be directly linked not only to physical problems like heart disease, but also to a lack of natural physical desire, a loss caused by romantic failures and trouble at work. José felt that this lack of desire was unmanly and understood ED treatment as an intervention that might restore the desires that would make him a normal man.

José stated that he had come to the IMSS for treatment: "Because I have an erection problem, I need stimulation, appetite to have sex with a partner. I've had this problem since 2003. I've lost the sexual appetite. I need to have—to be well." He reported that his "lack of desire" was linked to a series of problematic interactions with women, which in turn were grounded in other life problems. His sexual difficulties began in the same year that he split up with his wife; he said, "There was a problem. We were splitting up; my sexual appetite was diminishing." His divorce was directly related to health problems that had limited his capacity to work and earn money. He said that he and his wife split up because of "a historical problem. We were good, but when I got a pacemaker in 2003—when I was born, there wasn't money; my parents didn't take me to a doctor, and they say my heart is very slow—my wife saw that I wasn't going to achieve anything because I was sick."

At the time he received the pacemaker, José was working in construction and his wife was a housewife. He took a less physically taxing job, doing maintenance in a condominium community, but his wife complained that he did not earn enough and worked too many hours. As a result of these quarrels, he said, "Bad moods arose; we got angry over anything. We split." His wife stayed in their home with their three children, and he "went to the street." Soon after, José said that a final, emotionally problematic interaction with a woman cemented his sexual difficulties. He said that his sexual problems were caused by the trauma of his divorce as well as by this subsequent experience: "It was also due to . . . after that [the divorce], I met a girl; we tried to have sex. I saw the problem. She had an infection, she told me before . . . Herpes. That, when it got in here [pointing to his head], it really made me think, 'I shouldn't do this, how foul.' I didn't want to have sex."

However, he felt obligated to sleep with her, in order to comply with his manly duties and so that she would not be hurt. Nevertheless, he found it "difficult to achieve erection" and told her he could not do it again; "she left sad." José said, "After that, now I can't do it. If I see a pretty girl, I don't grow. I don't feel arousal; it doesn't appeal to me. I need to feel arousal, appetite. I don't feel anything. I don't think about it anymore—if I think about it, I get nervous."

While José saw his lack of sexual desire as rooted in multiple social causes, he also used a medical logic to understand it. Although even thinking about sex had become a source of emotional pain for José, he was seeking medical treatment that he hoped would restore his sexual desire and erectile function. He explained, "I want to return to having that, because it's normal, the normal life of a man. I think, I'm sick. I have a problem. I'm not living normally." He said that he wanted to have a partner again, although he could not meet one now: "I'm afraid that I won't function. Before, when I saw a pretty girl, lots of erection, arousal. Now even if I see a pretty girl—when I see someone I like, I get nervous, weak." It was clear that José hoped to return to what he considered a "normal" masculine state, in both mind and body, by overcoming the lack of desire caused by problems in multiple areas of his life. He crafted a hybrid understanding of his decreased sexual desire by attributing it to social and emotional causes, but also by viewing it as a bodily abnormality that could be treated medically.

However, José also subscribed to the idea that medically "forcing" one's body to perform sexually was a dangerous departure from its "natural" state. During his interview, José was holding a new prescription for the ED drug Patrex. When asked if he had been prescribed the drug, he said yes, but that he was afraid that it might be dangerous for someone with a heart condition. After this statement, he rooted through his medical file and brought out an old prescription for Levitra that his cardiologist had given him. He then pulled a somewhat crumpled box of Levitra out of his backpack, saying that he had filled the prescription but had not yet tried the drug. He said that he was hesitant because he tried Patrex before and was unsatisfied with the experience. He used it with a friend from his hometown, with whom he discussed his heart problems. He said, "I achieved erection, but very little, successful, but only a little. I was very tense, nervous. I had penetration, but was not that firm." José found it difficult to reconcile medical views of sexual function, the emotional pain and fear he now associated with sex, and the idea that ED drugs could be dangerous.

José's case reveals the fluidity of men's hybrid understandings of male sexuality. His erectile function, ideas about medical treatment, and experience of masculinity have changed along with his social situation and bodily health. He draws from different ideas about "natural" male sexuality to inform his own case and to make sense of the emotional pain he feels about his unsuccessful marriage and sexual encounters. He considers and attempts to use medical cures while simultaneously fearing that they may do physical harm. Similarly, José believes that his lack of desire is unmanly as it diverges from the idea of the unending sexual cravings naturalized by the discourses of machismo and the logic of medical ED treatment. He simultaneously rejects many elements of machismo, for instance, by seeking out tender relationships with women, and of the medical discourse of ED through his ambivalence about medical treatment.

Conclusion

Different ways of understanding decreasing erectile function and relating to its treatment are made possible by men's hybridization of different culturally available discourses about Mexican male sexuality: a natural bodily function that can become diseased, an innate drive definitive of Mexican manhood, an element of youth to be outgrown, or a practice contingent on masculine self-esteem and good romantic relationships. These different discourses all involve totalizing claims of what men, male sexuality, and decreasing erectile function naturally are. However, when men and their doctors fuse these discourses in order to make sense of their own embodied experience, these seemingly definitive statements about the nature of male sexuality are revealed to be building blocks which they rearrange to understand their lived experiences of sexuality and manhood. While the sorts of hybrid discourses that patients and doctors construct are constrained by structural factors and the limits of what makes sense in cultural context, they demonstrate that people's use of sweeping discourses about the nature of bodies, gender, and health does not translate into simple acceptance of these understandings. Thus, rather than transmitting particular norms of masculinity and health as they are sold around the world, or even being redefined in stable ways at specific local sites, medical interventions like ED treatment are woven into the flexible, ever-changing, hybrid understandings of selves and bodies through which people make sense of their lives.

NOTES

1. Pfizer is the drug company that sells Viagra.
2. José is a pseudonym.

REFERENCES

Amuchástegui Herrera, A. 1998. Virginity in Mexico: The role of competing discourses of sexuality in personal experience. *Reproductive Health Matters* 6 (12): 105–115.

Appadurai, A. 1996. *Modernity at large: Cultural dimensions of globalization.* Minneapolis: University of Minnesota Press.

Baglia, J. 2005. *The Viagra ad venture: Masculinity, media, and the performance of sexual health.* New York: Peter Lang.

Bass, B. A. 2001. The sexual performance perfection industry and the medicalization of male sexuality. *Family Journal* 9 (3): 337–340.

Becker, G., and R. D. Nachtigall. 1992. Eager for medicalisation: The social production of infertility as a disease. *Sociology of Health and Illness* 14 (4): 456–471.

Carlos, Y. 2009. Origen de la seguridad social en México. Monografias.com. http://www.monografias.com/trabajos11/imseg/imseg.shtml#EVOLU (accessed August 20, 2009).

Clarke, A. E., L. Mamo, J. R. Fishman, J. K. Shim, and J. R. Fosket. 2003. Biomedicalization: Technoscientific transformations of health, illness, and U.S. biomedicine. *American Sociological Review* 68 (2): 161–194.

CNN. 2008. Elderly men to get free Viagra in Mexico City. http://edition.cnn.com/2008/WORLD/americas/11/14/mexico.city.viagra/ (accessed August 20, 2009).

Daniulaityte, R. 2004. Making sense of diabetes: Cultural models, gender, and individual adjustment to Type 2 diabetes in a Mexican community. *Social Science and Medicine* 59:1899–1912.

García Canclini, N. 1995. *Hybrid cultures: Strategies for entering and leaving modernity.* Translated by C. L. Chiappari and S. L. López. Minneapolis: University of Minnesota Press.

González-López, G. 2005. *Erotic journeys: Mexican immigrants and their sex lives.* Berkeley: University of California Press.

Gutmann, Matthew C. 1996. *The meanings of macho: Being a man in Mexico City.* Berkeley: University of California Press.

Hewitt de Alcántara, C. 1984. *Anthropological perspectives on rural Mexico.* London: Routledge & Keegan Paul.

Hirsch, J. 2003. *A courtship after marriage: Sexuality and love in Mexican transnational families.* Berkeley: University of California Press.

INEGI. 2005. Causas seleccionadas de mortalidad por sexo 2005. In *Estadísticas vitales 2005.* Mexico City: INEGI.

Katz, S., and B. Marshall. 2002. New sex for old: Lifestyle, consumerism, and the ethics of aging well. *Journal of Aging Studies* 17 (1): 3–16.

Lizza, E. F., and R. C. Rosen. 1999. Definition and classification of erectile dysfunction: Report of the Nomenclature Committee of the International Society of Impotence Research. *International Journal of Impotence Research* 11:141–143.

Loe, M. 2004. *The rise of Viagra: How the little blue pill changed sex in America.* New York: New York University Press.

Mamo, L., and J. Fishman. 2001. Potency in all the right places: Viagra as a technology of the gendered body. *Body and Society* 7 (4): 13–35.

Marshall, B. L. 2007. Climacteric redux? (Re)medicalizing the male menopause. *Men and Masculinities* 9:509–529.

Mercado-Martinez, F. J., and I. M. Ramos-Herrera. 2002. Diabetes: The layperson's theories of causality. *Qualitative Health Research* 12:792–805.

Nichter, M., and N. Vuckovic. 1994. Agenda for an anthropology of pharmaceutical practice. *Social Science and Medicine* 39 (11): 1509–1525.

Ong, A., and S. J. Collier. 2006. Global assemblages, anthropological problems. In *Global assemblages: Technology, politics, and ethics as anthropological problems*, edited by A. Ong and S. J. Collier, 3–21. Malden, MA: Blackwell Publishing.

Paz, O. [1961] 1985. *The labyrinth of solitude and other writings.* Translated by L. Kemp. New York: Grove Weidenfeld.

Petryna, A., and A. Kleinman. 2006. The pharmaceutical nexus. In *Global pharmaceuticals: Ethics, markets, practices*, edited by A. Petryna, A. Lakoff, and A. Kleinman. Durham, NC: Duke University Press.

Potts, A., V. Grace, N. Gavey, and T. Vares. 2004. "Viagra stories": Challenging "erectile dysfunction." *Social Science and Medicine* 59:489–499.

Ramirez, J. 2009. *Against machismo: Young adult voices in Mexico City.* New York: Berghahn Books.

Rubin, R. 2004. Men talking about Viagra: An exploratory study with focus groups. *Men and Masculinities* 7 (1): 22–30.

Szymczak, J. E., and P. Conrad. 2006. Medicalizing the aging male body: Andropause and baldness. In *Medicalized masculinities*, edited by D. Rosenfeld and C. A. Faircloth, 89–111. Philadelphia: Temple University Press.

Tiefer, L. 1994. The medicalization of impotence: Normalizing phallocentrism. *Gender and Society* 8 (3): 363–377.

———. 1995. *Sex is not a natural act and other essays.* Boulder: Westview Press.

———. 1996. The medicalization of sexuality: Conceptual, normative, and professional issues. *Annual Review of Sex Research* 7:252–282.

van der Geest, S., and S. R. Whyte. 1989. The charm of medicines: Metaphors and metonyms. *Medical Anthropology Quarterly* 3 (4): 345–367.

Wentzell, E., and J. Salmerón. 2009. Prevalence of erectile dysfunction and its treatment in a Mexican population: Distinguishing between erectile function change and dysfunction. *Journal of Men's Health* 6 (1): 56–62.

CONTRIBUTORS

Shirlene Badger holds a PhD in sociology from Cambridge University. She is currently a research fellow at Anglia Ruskin University, Cambridge, United Kingdom. Her doctoral work explored what happens when a group of children who are severely obese and their families participate in a "genetics of obesity" study and potentially receive a genetic diagnosis for obesity. Currently, she is conducting research that continues to explore the experiences of obesity and other health conditions, with a particular interest in embodiment, diagnosis, and measurement, and the recruitment of children for complex interventions and medical/scientific studies.

Lynda Birke is visiting professor of anthrozoology in the Department of Biology at the University of Chester, United Kingdom. She is a feminist biologist who has worked for many years in women's studies and science studies and has been active in the women's movement. She started her career (and PhD) in biological science, researching in animal behavior for several years. She has published extensively in multiple fields and is, particularly, internationally known for her work in feminist science studies. More recently, she has focused on human/animal relationships, particularly in the context of scientific practice. She recently published *The Sacrifice: How Scientists and Animals Transform Each Other* (with Arnie Arluke and Mike Michael).

K. Smilla Ebeling studied biology and science and technology studies at the Universities of Hamburg and Bielefeld (Germany). She received her doctorate in 2001 in the history of science at the Technical University Carolo-Wilhelmina (Germany). She had additional education and appointments at universities in Minnesota, Hanover (Germany), and Basel (Switzerland). At the Carl von Ossietzky University of Oldenburg (Germany), Ebeling was a junior professor, served as director of the Center for Interdisciplinary Research on Women and Gender, and then was in the School for Linguistics and Cultural Studies. Her research focus involves animal studies, gender and science studies, and social science studies. Ebeling's most recent project is on gender and sexuality politics in zoos and natural history museums. She is currently an independent scholar in Hamburg.

Jill A. Fisher holds a PhD in science and technology studies and is an assistant professor in the Center for Biomedical Ethics and Society at Vanderbilt University. Her current research focuses on pharmaceutical clinical trials in the United States. Specifically, her publications analyze the new relationships among researchers, clinicians, and patients as well as the new ethical concerns and dilemmas that accompany drug studies conducted in the private sector. Her book on the topic, *Medical Research for Hire: The Political Economy of Pharmaceutical Clinical Trials* (2009), is part of Rutgers University Press's Critical Issues in Health and Medicine series.

Sara Giordano completed her PhD in neuroscience at Emory University. She has since received additional training in bioethics through courses and through working for two years on issues related to public health genomics and research. She is currently a postdoctoral fellow at Emory University, working on developing a bench-side ethics and community-based, participatory, research-training program in synthetic biology. Giordano is interested in feminist science studies, democratization of science, and questions of scientific accountability, more generally.

Jessica D. Horowitz received her bachelor's degree in women's studies and psychology from the University at Albany, New York. She completed her master's degree in women's studies there in December 2010.

Sel J. Hwahng, PhD, is currently a visiting scholar and adjunct professor at the Center for the Study of Ethnicity and Race at Columbia University and was a research investigator on the New York Transgender Project at the Institute for Treatment and Services Research, National Development and Research Institutes, Incorporated. Dr. Hwahng has received several awards, grants, and fellowships, including an Excellence in Abstract Submission among All Presenters Award from the American Public Health Association (HIV/AIDS section), an Independent Research Investigator Development Award from the National Institute on Drug Abuse, and a National Institutes of Health National Service Research Award. Publications include over seventeen articles and book chapters in peer-reviewed journals and edited volumes.

Iain Morland, PhD, is lecturer in English literature and cultural criticism at Cardiff University, United Kingdom. His research interests are the body, in particular, gender, sexuality, and body modification; cultural theory; and modern and postmodern theories of the self. He has published widely on the ethics, politics, and psychology of intersex in interdisciplinary books and journals. Publications include *Queer Theory* (2005), co-edited with Annabelle Willox, and a special issue of *GLQ* titled *Intersex and After* (2009).

Lesley J. Rogers is emeritus professor at the University of New England, Australia, where she coordinates the Centre for Neuroscience and Animal Behaviour.

She has a Doctor of Philosophy and a Doctor of Science from the University of Sussex, United Kingdom. In 2000 she was elected as a fellow of the Australian Academy of Science. Her publications include fourteen books (including *Sexing the Brain* and, with G. Kaplan, *Gene Worship*) and well over two hundred scientific papers and book chapters, mainly in the field of brain development and hemispheric specialization. She has received a number of awards for excellence in research.

Bonnie B. Spanier received her doctorate in microbiology and molecular genetics at Harvard University and then received major grants in the molecular biology of animal viruses from the National Institutes of Health and the Bunting Institute of Radcliffe College. She turned from biology to feminist critiques of science (*Im/partial Science: Gender Ideology in Molecular Biology*, 1995) as a professor (and chair) in the Women's Studies Department at the University at Albany, New York. She has written on biological determinist claims about sex and sexual differences. Her recent scholarship connects her community activism about breast cancer (she is co-founder of CRAAB! and NYS Breast Cancer Network) with critiques of information about cancer. She recently retired but continues to teach about women, health, and the environment.

Heather Laine Talley is an assistant professor of sociology at Western Carolina University. She recently completed her PhD in sociology at Vanderbilt University. Her dissertation, "Face Work: Cultural, Technical, and Surgical Interventions for Facial 'Disfigurement,'" examined the social significance of faces through an exploration of face work, or the material and cultural practices aimed at "fixing the face." The project was structured around four case studies, including the ABC reality show *Extreme Makeover*, facial feminization surgery, the philanthropic medical organization Operation Smile, and cutting-edge biomedical technology face transplantation. Her work employs qualitative methodologies to theoretically query sociology of the body, science and technology studies, disability studies, and queer sociology.

Claudia Wassmann is a postdoctoral research fellow at the Max Planck Institute for Human Development in Berlin. She completed her history doctorate from the University of Chicago (2005) and did postdoctoral training as a Dewitt Stetten Jr. Memorial Fellow in the History of Biomedical Sciences and Technology at the National Institutes of Health (2005–2006), National Institute of Biomedical Imaging and Bioengineering and Office of NIH History. She also holds an MD from the Free University of Berlin (1989) and a medical doctorate from the University of Düsseldorf (1991). Her research interests are in the history of science and medicine, the history of emotion, and biomedical imaging. She also authored documentary films for German Public Television and was a Knight Science Journalism Fellow at the Massachusetts Institute of Technology, 1998–1999.

Emily Wentzell is an assistant professor in the University of Iowa Department of Anthropology. Her research combines approaches from medical anthropology, gender studies, and science and technology studies to examine sexual health interventions' gendered social consequences. She is currently working on a manuscript based on ethnographic fieldwork in a Cuernavaca, Mexico, hospital, which examines older, working-class men's use and rejection of erectile dysfunction treatments and the influence of changing erectile function on their ideas about masculinity.

Angela Willey completed her PhD in women's studies as a fellow at the Bill and Carol Fox Center for Humanistic Inquiry at Emory University. Her dissertation research focused on the role of naturalizing discourse in the social, cultural, and historical constitution of compulsory monogamy. She is currently teaching as an LGBT studies postdoctoral fellow in women's and gender studies at Carleton College. Her research interests include feminist and queer theories; feminist science studies; feminist epistemologies; and history of race, gender, and sexuality in science.

INDEX

CPSIA information can be obtained at www.ICGtesting.com
260675BV00002B/1/P

9 780813 550473